Presentations in English

Jaquie Mary Thomas

Contents

Vorwort

Präsentationen vor Publikum sind bereits in der eigenen Sprache eine Herausforderung. Ungleich stärker steigt der Adrenalinspiegel, wenn man das sichere Terrain seiner Muttersprache verlässt und auf Englisch präsentiert: Ist mein Englisch gut genug? Was, wenn ich nicht auf das richtige Wort komme? Was, wenn meine Zuhörer mich nicht verstehen – oder ich sie nicht? Das sind nur die ersten Hürden, die viele bei einem Vortrag auf Englisch sehen.

Mit diesem TaschenGuide werden Sie nicht nur diese Hürden überwinden. Sie finden Werkzeuge, mit denen Sie Ihren gesamten Vortrag meistern – von der Vorbereitung, über die Begrüßung, die Präsentation selbst, bis zur Verabschiedung des Publikums. Auch auf die Diskussion mit Ihren Zuhörern können Sie sich gezielt vorbereiten – und dadurch sicher in die Präsentation gehen. Ich zeige Ihnen, worauf es beim Vortragen auf Englisch ankommt, und stelle Ihnen zahlreiche sprachliche Techniken und Beispiele vor. Ganz nebenbei eignen Sie sich nützliche Sätze, Wendungen und Key Words an.

Die Sprache ist das eine, die Herkunft Ihres Publikums das andere – und nicht weniger wichtig. Denn die kulturellen Unterschiede in der Art zu kommunizieren, sind groß. Und der beste Vortrag nützt nichts, wenn man seine Zuhörer schon bei der Begrüßung vor den Kopf stößt. Ich zeige Ihnen, wie Sie sich auf Ihr internationales Publikum einstellen und es so für sich gewinnen.

Jaquie Mary Thomas, Dipl.-Sozialpädagogin

Preparation

Whether you have five minutes, five hours or five months to prepare, these are the most important points:

Developing an international viewpoint

You have to do a presentation in English, maybe abroad or in your home country. Your audience may be from another country, or from many. What will they be like? What will they expect?

Things can be different

You want to do a good (or at least reasonable) presentation. You know your own idea of a good presentation but what's their idea of a good presentation? From an international viewpoint, a great many things can be different to presenting to a "home" audience. These can include:

- timing,
- content detail,
- how people listen,
- how / if they ask questions,
- eye contact (if any),
- conversation making,
- clothing,
- body language or
- even the question of whether a presentation is at all suitable or if everybody should have a good discussion over a 3-hour lunch with a bottle of wine instead.

It's a question of culture. A question of the way things are done in that situation, in that place, with those people.

Example

 Chinese and other cultures, where the group is more important than the individual, may come to a presentation as a group of ten to twenty or more people, depending upon the importance to them. They may then be surprised to see only you (with maybe one or two colleagues) and wonder where the others are.

The best way you can develop an international viewpoint and avoid a lot of misunderstandings, is to always keep the following four points in mind.

1 Accept that differences exist!

Think perhaps of a bottle standing next to a glass. From your point of view, the bottle may be in front of the glass. From someone on the other side it is behind the glass. From another person's viewpoint it's on the right. And, yes, from another person's viewpoint it's on the left.

Everybody can have a different viewpoint, a different view, but everybody can be right. Compare this to different places in the world. Presentations are done differently in France, in Germany, in Japan, in the US. Each of these ways of doing presentations is right, in those places, in those situations, with those people.

Examples: Starting a presentation

 Presentations in the US often start with a joke, in the UK with an apology, in Germany with the background details.

This does not mean that you should start with a joke in the US or an apology in the UK. It does mean that you need to think about the differences. How does your style suit the setting you are dealing with? You can then decide if you shorten the introduction and / or reduce the number of slides in total. You may then need to be more prepared to "go with the flow" – letting your audience point you in the direction they want to go to with their questions.

2 Opposite behaviour may not mean opposite values

Realise that differences in behaviour and differences in the importance of values can lead to a lot of misunderstanding. Opposite behaviour does not necessarily mean opposite values.

Example: Direct and indirect speech

 Germans tend to speak directly: "You made a mistake." This direct behaviour is often based on values of openness, honesty and the desire not to waste the other person's time.

The British tend to speak indirectly: "It seems that something wasn't quite right": This indirect behaviour is often based on values of politeness and respect for the other person and the desire not to hurt the other person's feelings.

The Chinese may say nothing at all.

> The translation "opposite behaviour = opposite values" here often leads to Germans thinking that the Chinese and the British are not open, not honest and waste time and, conversely, to British and Chinese thinking that the Germans are impolite, lack respect for other people (arrogant!) and don't care if they hurt other people's feelings.

This means that you shouldn't always let your reactions be led by how you interpret specific behaviour. Keep more of an open mind during a multinational or international presentation.

3 Use cultural generalisations with care

A generalisation is something that applies to 55% or more of a group of people – not a stereotype (100% of the people in that group, all of the time) and not a prejudice (a stereotype plus positive or negative judgment). A group can be a nationality group (e.g. French, Italian, Japanese), a regional group (e.g. north German, south German), a gender group (men, women), a professional group (IT people, sales people, commercial people, marketing people), as well as corporate, religious or age.

Example

"All French people interrupt presentations to ask questions" is a stereotype. "All French people interrupt presentations to ask questions – this is very rude" is a prejudice. The French people that you present to may not interrupt at all, and if they do, they will probably see it as positive.

You can use generalisations to help you prepare for your presentation, if you think of them as "most of the people do this most of the time". So you can expect interruptions from Italians in general or active listening (nodding, verbal agreement or disagreement) from Americans in general but, if the people you deal with don't do that, then you shouldn't be too surprised. Everyone is an individual, with their own unique history – you can only identify tendencies that may or may not apply.

4 When and how to adapt to others' cultural style?

People often ask "Why should I change, why don't they?" and may feel that they are not being "themselves" if they change their style.

First and foremost, we all change our style depending on whether we are talking to our colleagues, our managers, our partners, our friends, our children, etc. It doesn't mean we are not being "ourselves" if we change style. We just have different roles.

This is the same in business. Sometimes you present to your colleagues, sometimes to your managers, maybe to your team members, as well as suppliers, customers, collaboration partners – again you have different roles and adapt often quite naturally, without thinking about it too much. Presenting in English can mean you are presenting to any of those just mentioned – maybe two nationalities, maybe twenty nationalities, maybe in your own home country, maybe abroad. All of these situations are different and, if you want to do a good presentation every time, you may need to adapt in every situation.

When to adapt: situation, surrounding and individuals

This decision of when to adapt to others' cultural style can only be made by you (or you and your colleagues when working as a group). There are no clear black and white guidelines, but you can make the decision easier by considering the situation, the surrounding and the individual.

- **situation:** This is your first factor when deciding to adapt. What is the business situation and what does the distribution of power look like? Are you making your presentation as customer or supplier? Are you an engineer presenting technical information to a potential buyer or transferring information to other people on your international project team? Are you a financial controller presenting to your board or to your team? People are generally more motivated to switch styles when they need something.

- **surrounding:** Where are you? This is your second factor when deciding when to adapt. Are you presenting in your home country or abroad? Are you in your own company or in theirs? Are you in the boardroom or on the factory floor?

- **individuals:** Who are you presenting to? This is your third factor when deciding when to adapt. Who are the people you are dealing with? Are they men, women, older or

younger than you? Are they board members, sales or marketing people, human resource experts, accountants, engineers? Do they have a lot of international experience?

How to adapt: cultural dimensions

If you don't know or are unsure about any of the above three
points – e.g. who you are dealing with – then you need to do
some thinking (see "Put yourself in your audience's shoes",
page 25). Once you have made the decision of **when** to
adapt, you need to think about **how** to switch style. You need
to consider the possible cultural influences regarding the
situation, surroundings and individuals according to cultural
dimensions. When you present internationally, the following
cultural dimensions are usually the most important:

- **time orientation** – how is time likely to be seen? Could
  your audience be more, the same or less punctual than
  you? Do you think they see time as money or do they treat
  it as something that will continue endlessly? Intercultural
  researchers talk about monochronic and polychronic cultures. People belonging to monochronic cultures prefer to
  do one thing at a time and place importance on deadlines
  and time schedules. You need to be aware of your natural
  style in comparison to the people you are dealing with.
  Then you can decide how to switch style by becoming
  more or less punctual, for example.

Example

When planning a presentation to a Spanish audience, you may find that they could probably be less punctual than yourself, and that they don't cut time into specific numbered pieces as much as you do.

This means that for a presentation scheduled to start at 09.00, you don't need to worry if half of the people are not there and you start later than originally scheduled. Also, that your presentation may go well over schedule. This may not matter to them as much as it does to you – again no need to worry.

- **People vs. task orientation** – how important to your audience is the personal relationship compared to the task? Do they split or mix the two more or less than you? In Germany, people tend to separate business and pleasure – "Dienst ist Dienst und Schnaps ist Schnaps". Germans generally do the business first, and then maybe get to know the people. Other nationality cultures, such as the Japanese, will want to get to know the people before doing business.

Example

When presenting to an audience more person-oriented than yourself, make sure you allow time for relationship building – getting to know each other – e.g. dinner the evening before the presentation or coffee and cookies before you start.

- **Communication style – direct vs. indirect** – how directly or indirectly does your audience communicate? Again, you need to be aware of your natural style, so that you can choose to "soften" your English language with phrases

such as "I think", or "perhaps" or "maybe", as well as opposites "not easy" instead of "difficult", or rephrasing
"challenge" instead of "problem".

Example

 When formulating objectives, don't say "you will ..." but "I hope
you will ...". For example, "By the end of this presentation, I
hope you will have a clear picture of progress on the project."

See the international viewpoint preparation checklist on
page 28 for further details.

Preparing yourself, the person

Most people have nerves before making a presentation – in
fact, you're unusual if you don't. On top of any normal nerves
comes the fact that you need to do a presentation in a foreign language – English. Finally the uncertainty about the
international audience: How will they react?

How to deal with nerves

First, you need to fundamentally accept how much you like
making presentations or not. That's the way it is. You need to
remember that you are an expert in your specific field. Communicating specific information to others is part of your job.
Doing that well means doing part of your job well. You have
probably already given other presentations and they went
well enough. Or at least, you didn't lose your job.

Three golden principles to remember

1 **You don't need to be perfect to succeed.**
   You can be average, or below average, you can make mistakes, too. As long as you give the audience something of value, it doesn't matter. They will be thankful if they walk away with something of value.

2 **Your audience is usually not sitting and waiting for you to fail.**
   Most of them are scared to death of public speaking. A slip of the tongue or a mistake of any kind may seem big to you, but it's not usually very meaningful or important to your audience. Their judgments of you will be much more easy-going than your own. It's useful to remind yourself of this point, especially if you think you performed poorly.

3 **You are unlikely to please 100% of the audience.**
   That's okay. Even if you would like to please everybody, no matter how good a job you do, there is likely to be at least one person in your audience who will disapprove of you or your content. That's human nature!

Think back to the last few presentations you watched. Were the mistakes that the presenters made important? Do you even remember any mistakes?

Techniques to reduce some of your nerves

Choose between one and three of these techniques and use them for your next presentation. If they don't work for you, then try others. It's unlikely that you will be able to eliminate your nerves completely – indeed, having some nerves is positive, providing you with the necessary adrenalin.

- Make sure you have the key English words for each slide either in the slide itself, or on numbered cards. Have one or more solutions ready, if you forget the words (see "What if you forget the words?", page 20).

- Learn the first few sentences and last few sentences of your presentation word for word.

- Visualise yourself starting your presentation well – close your eyes and see the first few steps – e.g. walking to the front, calmly setting up your laptop, checking the image, turning to the people, taking a deep breath, saying good morning and outlining the purpose of the talk today

- Visualise yourself ending your presentation well – close your eyes and see the final few steps – e.g. looking at the audience and thanking them slowly and clearly, gathering your papers / equipment calmly and returning to your place.

- Remember your audience are people – just like you, they eat, breathe and sleep. No more, no less. Just people.

- Remember past experience – somehow, somewhere, you've been through this type of thing before. You know this feeling, you got through it and you survived. You can do it again.

- Let go of "I can't" – research shows that if you focus on "I can't do it", you reduce your chance of success. Logical really, isn't it? Stop telling yourself you can't do it and focus on doing things differently. That's one of the reasons you're reading this book.

- Just do it! – no comment needed here.

Dealing with "language" nerves

First, you need to fundamentally accept that you have to do the presentation in English. Even if you don't like it, even if you wish that you could do it in your own native language. There's probably nothing you can do about it, so just getting on with it is the easier policy.

"My English isn't good enough.", "Everyone else speaks better English than me.", "What if I forget the words?" and "What if I can't understand them?" are typical thoughts running through people's minds.

Your English isn't good enough?

Second, you need to accept however good or bad your English is. That's the way it is at the moment. Most non-native speakers always feel that their English is not good enough. Whether they are at an advanced, intermediate or elementary level, their attitude is still the same – not good enough.

People would like their English to be as good as their mother tongue. This is unrealistic and illogical.

Just think of the number of conversations you have had, the number of books and newspapers you have read, the number of television programmes you have seen, the amount of work you have done in your mother tongue and compare it to the same in English. Don't create stress for yourself with negative guilt feelings that you really should attend an English course, learn more vocabulary, practise more, etc. Work and information flows have become much faster over the last couple of decades. Business trips by air, leaving and returning the same day are commonplace. Internet information sourcing means people process huge amounts of information very quickly. Process steps are performed much more quickly than before via e-mail, Netmeeting and other electronic methods of communication. All of these put demands on people's time. They are often left with the feeling of not having enough free time. Therefore, if you do manage to attend an English course, that's fine. If not, don't worry about it. Just get on with the job.

Everyone else speaks better English than you?

Next, think of your audience. If you have non-native speakers, it is highly unlikely that they will notice your mistakes. It could even be that their English is not as good as yours and you may need to simplify your English (use shorter, not longer words) and speak more slowly than usual (i.e. with pauses between phrases). If you have native speakers, they are probably impressed with the fact that you can speak a foreign language. The majority of native English speakers cannot speak a second language. Switch your attitude and try to be proud of the fact that you can speak English (if you aren't already proud).

If you think that your colleagues speak better English than you, remember again that you, too, are an expert in your field. Don't hesitate to consult them whenever a word is missing (see next section, page 20).

Don't feel inferior language-wise if presenting to Scandinavians, a great many of whom can speak fluent English. This results from the fact that their television films are shown in English, with subtitles in their native languages. They have a huge advantage. Try and put them in the category of native speakers.

> Listeners absorb information differently to readers. You need to speak using the "spoken word" – not the formal more complicated "written word".

Checklist: dealing with "language" nerves

- Make your presentation Concise, Clear and Complete (C.C.C.).
- Use short sentences.
- Use simple words.
- Avoid phrases that are difficult to say.

What if you forget the words?

A person forgetting their words completely is extremely unusual. Truly unusual. The chances are minimal. If, however, you are one of the people who have the experience of drying up completely in the past and not being able to say a word, well, even if it was horribly embarrassing, we are all human at the end of the day and the audience most probably sympathised with you. One of the best techniques is to learn the first and last few sentences of your presentation by heart. Have it checked, if you have time, by a native speaker (colleague, another department, friend) and then learn it so that you can say it off by heart. This means you have to practise saying it. Not just in your head. Do make sure you speak the words aloud, again and again, in the car, on the train or bus, in the bathroom, until you know them inside out. It is more likely that you will not be able to think of a specific word(s) during the presentation. You have a selection of alternatives:

Solution 1

If other people in your audience are native speakers of the same language as you, then the immediate solution is to say:

- I'm looking for the word "zeitraubend" in English.
- What's "zeitraubend" in English?

Either someone will give you the answer and say "time-consuming" or will try and explain "takes a long time", or nobody will be able to help you. At the very least, you will not be alone. This solution really works very well and needn't interrupt the flow of your presentation.

Solution 2

If the word doesn't come to mind, close your mouth (to prevent an "err", "umm" or "aah") and just give yourself a little time to find the word or rephrase or start another sentence. Most people speak too quickly during presentations and your audience will welcome a break to think or to let your last point sink in or to even just relax for a moment or two. Don't be afraid of a silence.

Solution 3

If you can't find the word and have no native speakers of the same language as you, try involving the audience:

Example

You: This new process is much more ... what's the word? It won't break down so often, it's much more ...

Audience: Safe?

You: Not exactly

Audience: Reliable?

You: Yes, exactly, thank you – it's much more reliable than the old system.

You can substitute "What's the word?" with "I'm looking for a word ..." or "I can't think of the word ..." If the audience doesn't know the word either, stay calm and move on to the next sentence. Use the words "anyway" and / or "well" to link or thank the audience:

- Anyway, what I want to say is that it really is a very good system.
- Well, the main point is that this system really is very good.

What if you can't understand the audience?

People often worry that they won't be able to understand the audience. English, Scottish, Welsh, Irish, west coast American, east coast American, Australian, etc. all have extremely different accents. It's not uncommon for native speakers to misunderstand each other. Again, the key is to stay calm and get others people's help if necessary. Don't appear aggressive by telling someone to speak more clearly – always take the problem away from the speaker.

Useful phrases

- *Acoustic problem:* Sorry? I couldn't hear that. Could you say it again a little louder please?
- *Too fast:* Sorry? I'm afraid that was a bit too fast for me. Could you say it again a little slower please?
- *You didn't understand:* Sorry? I didn't get that. Could you say it again?

Then check by repeating the statement / question:

- You're saying / asking ...
- What you're saying / asking is ...
- What you want to know is ...

Example

A: What do you think it all boils down to in the end? Are there really any significant differences between all three options at the end of the day?

B: Sorry? I didn't quite get that. Could you say it again?

A: Sure. What I want to know is whether it really makes a difference which option we take here, or if they are all pretty much the same.

B: So you're asking about the differences between the three options?

A: Yes, that's right.

- If you still don't understand, try to get the audience to help: I'm sorry, I still don't quite understand. Perhaps someone could translate for me?

- Or ask the person if you could talk together in the break or at the end, depending on the situation. Be specific – "in ten minutes" or "at 10.30" – otherwise you may sound like you don't mean it: "I'm sorry I don't understand right now. Can we meet in the break in ten minutes and talk together? I'd like to take some more questions before we finish."

- Another solution could be to use a flipchart to visualize the statement / question, if suitable: "Let me put that on paper, to be sure it's clear. Your idea is to do A first, then B, followed by C and D. Is that right?"

In situations where it is important not to lose face (e.g. Japan and most Asian countries in general), it may be a good idea to give an answer (some general comments) even if you don't understand the question. In this way, the questioner is not embarrassed, you don't lose face either and you can always speak to the questioner.

See the section on dealing with questions (page 87) for more details.

Understanding non-native English speakers better

It's often difficult for native English speakers to understand each other. The differences between American, Australian, British and Indian English are huge. Even differences within a country (Oxford English and Scottish English, for example) make things difficult. The next two points can be helpful.

- A lot of Asian languages are made up of "consonant, vowel" pairs – e.g. "Mi – tsu – bi – shi." This often makes it difficult to say consonants at the end of English words – e.g. "foo" could be food, fool or foot, which are all "consonant, vowel, consonant" patterns. Listening is made easier if you know this fact. Speaking is made clearer by adding a "weak" vowel after the final consonant – e.g. "foo – t"

- Indian people in general often speak fast, in a sing-song voice without pauses. Either you get used to it within a couple of days, or you "gently force" the person to make breaks. Interrupt carefully by lifting your hand(s) up and repeating the keywords of their last phrase then put your hands down to "let" them carry on. If you know them well enough, you could ask them to make pauses.

Putting yourself in your audience's shoes

Try to see things from your audience's point of view. Not yours. This is vitally important – whether for 6, 60 or 600 people. "What do **they** want to know?" is usually different to "What do I want to tell them?" You cannot start to prepare a presentation without having at least some idea of who your audience is. If you start preparing before you know who they are, it means that you are telling them things from your polnt of view only. Not theirs.

Imagine you are sitting in the audience, watching and listening to your presentation. In your mind's eye, go and sit in a chair and be part of the audience:

1 Who are you?

2 Why are you there?

3 What do / don't you want to know?

If you don't already know the answers, even the smallest amount of research or thinking will show large benefits. Ask the person who requested you to do the presentation. Ask your colleague who was with the same people last month. Ring and ask a department secretary. Find out the following: Are they younger or older than you? What are their jobs? What are their management levels? Are they more or less technically-minded, sales-oriented, commercially-minded than you? You may need to acknowledge these points.

Examples

If presenting to older people, recognise their experience to start with:

"Good morning everybody. It's good to be here in our Berlin office this sunny morning. For those who don't know me already, my name is James Harrison from the Frankfurt sales office. During the next thirty minutes, I'd like you to see how we gained a new key customer last month. I do realise that a lot of the people in this room today have a great number of years' experience in this field already. The idea is to share this case study with you."

If presenting to a multi-disciplinary audience, recognise this fact, too:

"During the next thirty minutes, I'd like you to obtain a basic overview of the new process you will start using this June. I know that we have technical, financial, sales, legal and quality assurance people in this room today. That means that some parts of the presentation will be more interesting than others depending on who you are. The main point today is that everybody has a basic overview – specialised training for each department will take place next week."

You need to formulate objectives and plan content for the presentation, based on your audience's needs as well as yours. Your audience's interests can be very different.

Examples

1 Who are you?

A: a board member, listening to the fourth presentation from a project manager that day (27th that month)

B: engineers on your project team who have a lot of other meetings booked this week.

C: company trainees

2 **Why are you there?**

A: here to make sure the project is going well.

B: here to be updated on the project progress.

C: here because the overview of your work in your department is part of the training course.

3 **What do / don't you want to know?**

A: wants to know the main points of how the project is going, including how difficulties are being dealt with. Doesn't want to have to "search" for the difficulties. Doesn't want to waste time. An interesting real-life story might be of interest.

B: wants to know only your main points, especially any potential interfaces with their work. Doesn't want too much irrelevant background detail.

C: wants to hear what they could really be doing in your department. Wants to know the most interesting and most boring parts of the work.

- Use "you" or "we" but not "I" (more interesting): "You will be able to see." not "I will show you."

- Use "need" but not "must" (very direct, stressful): "Today we need to plan a rough time schedule." not "Today we must plan a rough time schedule."

Useful phrases

- In the next twenty minutes, you should get a good idea of the progress made in the main areas of the project, plus an overview of how we are dealing with current difficulties.

- So, here we are again for you to get the latest update on our project, specifically concentrating on any areas of overlap.

- Our main aim today is to compare current and projected performance with the targets. I can tell you now, fourteen out of twenty-six regions are on or over target and twelve are either just under or struggling. The total balance is slightly less than target, although 3% up on last year. Let's start with ...

Checklist: preparation international viewpoint

If you have to present to people from only a couple of other countries then you can research as follows:

- Find a cultural insider who can give you main "do's and don'ts" – a person from that country who now lives in your country or works in your company or a contact you already have abroad or a colleague who was there before.

- Consider the cultural dimensions mentioned in chapter one on page 12 (time, person vs. task orientation, etc.).

- Don't forget to ask yourself, when dealing with a more person-oriented culture, whether a presentation is really suitable? Should you be starting with dinner the night before? Or planning a long lunch? Or is a factory tour followed by lunch / dinner more appropriate?

- Search for "How to do business in ..." on the Internet.

- Buy a book.

If you have a multi-national audience from many different countries:

- Consider the cultural dimensions as above.

- Make sure you have both a clear and flexible starting time! For example, 08.45 Coffee and greetings, 09.15 First talk.

- Make sure you have both a clear and flexible agenda – for example, five points written in a circle with a total start and finish time, clear for the monochronic people, with the option to change the order for the polychronic people.

Organising facilities

If you're going to make a presentation in another country, be prepared to be flexible. Don't expect to always have the same standards everywhere.

Be prepared

- Have a backup of your PowerPoint presentation on a (second) USB stick or even printed out on overhead projector transparencies, depending on where you are going. Carry it with you in a different place (not both in your briefcase).

- Consider the power supply and take your own international adaptor or two (usually on sale at airports).

- Take your own pens with you, if you want to use a flipchart or whiteboard. Better still, take your own flipchart paper and masking tape (plastic cylinder carriers are light and very cheap).

Organising the setup

How much of the setup you can influence depends, of course, on the situation and whether you are the customer or supplier, guest or host, etc. If you need to make the arrangements yourself, the main requirements you may need to consider are:

- room size and room set-up – U-form or hollow square for small groups, theatre or classroom style for large groups
- equipment – projector, flipcharts, pinboards
- acoustics – microphone, lectern
- refreshments – water, tea, coffee

Even if you clearly requested specific arrangements, these may not be considered suitable and the organisers may be culturally unable to refuse the request without offending.

Example

 Requesting a semi-circle of chairs with no tables for a training course in China may be pointless due to the necessity of tables for the supply of tea. You may not know this until you arrive, because the organisers are unlikely to want to disagree with you beforehand and may feel certain you will understand when you see that the tables are there.

Useful telephone phrases

- Good morning, this is Klaus Schmidt calling from ABC.
- I'm coming over to hold a presentation on the 15th of March and I need to speak to the person responsible for the organisation. Is that you?

- I just wanted to ask a few questions. Is now a good time? Okay. I understand that there will be about 15 people, is that right?

- And could you tell me how big the room is?

- Could we possibly have the chairs and tables in a U-shape?

- And do you have a projector I can attach to a computer or should I bring one myself?

- And is there a flipchart in the room?

- That sounds fine. And my final question is about refreshments. Will there be any drinks and snacks?

- How are the acoustics in the room? Will everyone be able to hear me, or should I have a microphone?

- Can you tell me wether here is a stage or a podium? And do you have a lectern or does the presenter move around?

- Could I have your e-mail address and then I can just send you a short e-mail to confirm?

- I think that's all for now. Can you let me know if anything changes?

- Okay then, thank you very much for your help.

"Beamer" in English means a BMW car. You need to use "projector" or "data projector".

E-mail examples

Re: Presentation, March 15, 10.30

Dear Ms Woods

With reference to the above presentation, I would appreciate it if you could organize the following:

- 15 places in a U-shape
- 1 data projector (to attach to my laptop computer)
- 1 flipchart plus pens
- refreshments for the break
- as well as a table for lunch in a nearby restaurant for 15 people at approx. 13.00.

Please let me know if the above is possible.

I look forward to hearing from you.

Best regards

Re: Presentation, March 15, 10.30

Dear Peter

It was good to talk to you this morning. I'd just like to confirm the requirements as follows:

- seating for 35-40 people
- 1 data projector (to attach to my laptop computer)
- refreshments for the break
- 45 hard copies of the presentation

I'll send you a soft copy of the presentation 2-3 days beforehand, as agreed.

Thank you for your help.

Best regards

You yourself may prefer e-mail, but a lot more person-oriented countries prefer to communicate by telephone. Doubling up (or even tripling) can be more effective – e.g. call the organiser in Greece, then confirm by e-mail and perhaps ring again the day before, depending on the situation.

Your presentation structure

Circular vs. linear structure

International presentations need to be both circular and linear, because some nationalities are polychronic and others monochronic. The following could happen to you when presenting to an international audience. You may plan carefully and logically according to your preferred system – first, background detail and history, then the problem itself, next, possible options and, finally, your recommendation. You find, however, that your audience wants to know your final recommendation first and asks you a lot of questions, forcing you to jump back and forth from point to point in a circular fashion. Maybe they're not interested in either the background or the details, but want to concentrate on the outcome.

Five golden structure rules

This means you should

- show an agenda slide but be ready to move between sections as required,

- provide background and details in the slides for the analysis-oriented people (e.g. Germans), but watch the audience's reactions and skip over them if you think suitable,

- not "hide" any recommendations at the end but state them right at the beginning, if possible. Example: "So, we need to find a solution to problem A today. I recommend B and want to show you what the problem was, the different options and why I recommend B, so that you can then make a decision.",

- definitely not say "I'd like to come to that point later" (that can seem very offensive) but give an answer "I think option B is better because of X, Y and Z. The problem we had was ..." and go back to where you were, or just spring forward and leave out the details,

- don't always ask your audience to keep questions to the end – but ask for questions during the presentation, in order to suit their style as necessary.

If you have your agenda written on a flipchart (which you can prepare beforehand and take with you, or printed cards / A4 sheets, with sticky tape), you can refer to it and people can see it all the time. This can help you and others keep track when needed.

Timing

If your presentation is scheduled to start at 10.00 and you expect it to start at 10.00, 10.01 or 10.02, latest 10.07, then you could get frustrated working internationally. Remember the most important thing – what is your overall aim for the audience? It's not to follow your time schedule. Most people try and pack too much into a presentation – whether for an international audience or not.

Allow breathing space

Most situations need a plan with a lot of breathing space because of the following three points:

- People might just not yet be there at the time you're due to start. This can be due to the concept of time which is viewed differently worldwide. Some cultures consider time to be a never-ending, constant quantity. "If we don't use the time and do it now, we can always do it later".

- Others view time as an event-related concept: if something more important happens, that takes priority. You need to expect this. "Feel" the situation – if it starts to feel uncomfortable, ask the person in charge if / how long you should wait.

- People might be drinking coffee and chatting, not ready to start. This can be due to different person / task orientation at work. Some cultures need to get to know the person first, before they can do the business.

Again, you can expect this from polychronic cultural groups. You need to "feel" the situation, go with the flow, depending on the distribution of power. Alternatively, plan coffee and informal introductions at the start, if you can. Don't become stressed about your time plan – remember your aim. People might want less input from you and more discussion. Again, adjust accordingly – remember your aim. Put "approx." (short for approximately) or "circa" in front of times, prepare an agenda with few details – that way you can be more flexible.

Example: agenda

 08.45 Coffee and refreshments, informal introductions
09.30 BSN Motorparts, James Holloway
circa 11.00 Break
BSN Motorparts (continued)
circa 12.30 Lunch
14.30 PCR Chemistry, Annette Rowlings
circa 16.00 Break
18.00 Close

Structure – main components

1 Introduction: Tell them the point and the plan (1-2 minutes max.).

2 Main body: Tell them the details, including any difficulties and proposals or next steps.

3 Summary: Tell or remind them of what you just said, including the point and next steps (should be shorter than introduction).

Questions are ideally taken all the way through, depending on the situation.

Structure in detail

Structure	Elements
introduction	
opening	• greeting and name (slowly and clearly) • company and / or department • reference to place and time (here and now)
objective	point of presentation (from audience point of view)
overview	one slide agenda (ideally three main points)
organisation	time available, questions welcome throughout or afterwards, refreshments, etc.
Main body	
starting signal	Say that you are starting: "So, let's start with ..."
three main sections	People remember three things best – split your presentation into three parts, with sub-divisions later if needed. For example: – new regulations in export paperwork – main differences to current paperwork (details section by section) – future references and assistance

realia	real-life relevant pictures of product, site, people and interesting stories
graphs, charts, diagrammes	Explain "big picture" first, then details second.
Summary (introduction in reverse)	
ending signal	Say that you are ending: "So, that brings us to the end."
review	Show agenda slide again.
objective achieved	Re-state objective: "So, I hope you now have a clear picture of ..."
open discussion	depending on time
next steps	State next steps or action required: "Our next meeting will be on ..."
thank the audience	Say: "Thank you everybody."
Question and answer	
	throughout your presentation or at end: ▪ listen ▪ pause ▪ repeat (part of) question ▪ answer ▪ link to a main point

How to prepare good slides

Most important of all – you really don't need to spend most of your time preparing perfect, detailed slides for an international audience. Remember that person orientation is likely to be at least, if not more important than task orientation. For the French and Belgians, for example, the objective is to be creative and provoking – make the audience think and reason. You need to know exactly the aim and the key messages of your presentation and be prepared to talk about those with the audience. Discussion and interaction are most usually more important than perfect slides.

Good slides are clear and to the point. They help show something to people, so that they can understand better. You need to ask yourself:

- What's the point of this slide?
- What do the people need to know?
- Can it be clearly read?

Checklist: preparing slides

- The slides in your international presentation need to be in "international English" – you should avoid long, complicated words.
- Always get the slides checked by a native speaker for spelling and grammar – at the very least, use the "spell-check" function.

- Make sure you have a note of the English keywords for the main message of every slide. Ask a colleague or use an online dictionary. Don't spend hours on it – stay realistic.

- Use the "six-pack" rule: six words per line, six lines per slide are ideal.

- Bullet points with key words are easier to read and remember than long texts.

- If your company working style means you must make long sentences, then don't pack them all together, but spread them out on more than one slide.

- Keep it simple – three colours maximum and the minimum of font styles (preferably one only) and sizes (preferably two, maximum three) make the slides much less confusing and much easier to read for your audience.

- Pictures speak one thousand words – use photographs of real products, systems, locations or people whenever possible, to bring your presentation to life and make it memorable.

German language speakers often use too many capital letters – especially in headings. Capital letters should be used in business English only at the beginning of sentences and for specific names, e.g. Reservation system overview, not Reservation System Overview.

Greetings and introductions

What you say and do when meeting the people and getting started with your presentation all counts towards the first impression.

This chapter will help you to find the right phrases and feel confident about making a good start:

- what to say when you enter (page 42),
- introducing your presentation well (page 45),
- introduction components (page 48),
- dealing with handouts (page 53),
- taking care of technical problems (page 53).

What to say when you enter

We all know that greetings are not the same the world over. If you choose to match style, points to watch for are:

- whether handshakes are made or not (less common after the first meeting)
- strength of handshake: you should mirror, strong with strong and gentle with gentle
- amount of eye contact or looking away: again, mirror – don't force someone to look in your eyes if they don't want to
- distance between persons: could be much further or closer than you are used to
- use of first or last names: this often causes uncertainty. Listen carefully and then match the style. Americans, Australians and British tend to use first names very soon, if not immediately – "You can call me Robert." Other cultures use "Mr" and "Ms" to start with and may or may not switch to first names later. Chinese often have an additional western first name, which you should then use. Japanese often add "san" at the end of the name
- conversation making: be prepared to do it!

Example

When doing business with the Far East, watch out for card exchanging when greeting. Koreans, for example, will stand apart, approach each other, bow, exchange business cards using both hands, step back and then read each other's business card aloud, with appropriate remarks "I see you are the Business

Director …". This shows respect for the persons and their positions. As a foreigner, you are not expected to know how exactly to bow (different depths show different levels of respect), but you should not write on the business card and definitely not put it into the back pocket of your trousers. Watch to see what the other does.

Meeting people for the first time

Example

"Hello, good morning. My name's Heinrich Melke. Nice to meet you." – "Hello. It's nice to meet you, too. I'm Alexander Thompson. Call me Alex, it's easier." – "Okay then Alex, I'm Heinrich, or Henry, whatever you prefer." – "Henry's easier for me!" – "I'm not surprised, Heinrich is a bit of a mouthful if you don't speak German." – "Yes, I'm always impressed by how well Germans speak English." – "Oh, it's not easy, you know. We are perfectionists – it's never good enough!"

"Hello, good afternoon. I'm Wolfgang Steinecke from ABC GmbH." – "Ah, yes, good afternoon. I'm Roger Silvestre and this is Sylvie Dalmar." – "Nice to meet you." – "Nice to meet you too." – "Did you find the way here okay?" – "Yes, it was no problem at all. It's a very nice place. Is it new?" – "Well…."

Useful phrases

- Hello, I'm Andrew, Andrew Smith. It's good / nice to meet you.

- Hi, I'm David Jones. (I'm) pleased to meet you, too.

- Hello, I'm James Morrison. It's good / nice to meet you, too.

- Hi, I'm Mary Thompson. Pleased to meet you, too.

"It's good / nice to meet you." and "(I'm) pleased to meet you" are interchangeable. "How do you do?" is rather more formal British English. It means "Pleased to meet you" and does not mean "How are you?" The answer is simply "How do you do" or "Nice / good / pleased to meet you."

Most people all over the world use "Good to **meet** you" for the first meeting, then say "Good to **see** you again" for further meetings. Americans, however, often use "Good to **see** you" for the first meeting, too.

Making conversation

Making conversation is often a very important part of the business. Try not to fire questions "bullet style" but say a little yourself first before asking.

Phrases like "It's a nice day, isn't it?" are usually not really invitations to talk about the weather, but an opening to see if the other person wants to talk. A suitable answer if you want to talk is "Yes, it's lovely. Much nicer than when I left home in Kiel, northern Germany yesterday. Where do you come from?"

If someone asks "How are you?", you should not only give an answer, but also ask back. You shouldn't directly translate into German and answer with a list of your latest illnesses, but think of it as meaning "Good morning".

> Avoid talking about politics, sex or religion.

Useful phrases

- So, here we are. This is my first trip to ... I have heard so much about it. Do you live nearby?

- I come from ..., myself. Have you ever been to Germany? (*British English*) / Did you ever go there? (*US English*)

- There are a lot of trees just starting to flower at the moment. Is it the beginning of the cherry blossom season?

- I'm really looking forward to trying the food / wine / beer here. I have heard so many good things about it.

Examples

 Good subjects for Americans can be sport – baseball, basketball or American football – ask about the local teams e.g. "What are the main sports around here?" Chinese people often love to talk about food e.g. "We have a lot of Chinese food in Germany, but I'm really looking forward to a real Chinese restaurant here in Beijing."

Introducing your presentation well

This is the start – get it right! A good guideline to make sure your presentation will be acceptable for the whole world is to make very sure you clearly (and quite slowly) state who you are and the point of your presentation, with some reference to "here and now" to make it "real" and interesting, together with a few details of the plan. The title and plan should be visualised in simple English – either on a slide or a flipchart (advantage – it remains in view).

Examples

Presentation 1

Good morning ladies and gentlemen, I'm Benno Donauer from the electronic engineering department at Opus GmbH and it's good to be here in Rio on this very sunny Friday morning. By the end of this presentation you should be able to see why A, B and C are so very important to you and your company, Omega. First we'll look at A, then move on to B and finish with C. We have forty minutes altogether. If you have any questions at any time, do please just ask.

Presentation 2

Good morning everybody!

For those who don't already know me, my name's Bernd Huber. I'm from the Electronics Assembly Systems Department at Drive Technologies AG in Berlin. Here we all are in Madrid, today, October 5th, 2008, and by the end of this presentation (*show title slide, read name of presentation*) "Serrano Limitido and Drive Technologies" I hope you will know how technologically advanced, reliable and flexible our machines are for you, and be able to see why we at Drive Technologies are the very best partner for you.

We'll look at the X-machine technical functionality to start, then move on to possible adaptations for you, Serrano Limitido, and finish with our service offer – how we can work well together.

We have three hours scheduled, with a break halfway through for coffee. If you have any questions at any time, please do ask me. Good. Let's get started.

Checklist: introducing your presentation

- Be sure to formulate your objective(s) carefully from your audience's point of view – who they are and what they need (see "Put yourself in your audience's shoes", page 25).

- If you know that problems are likely to arise, you can either talk to people about them beforehand or name them at the beginning of your presentation.

  – "So, we are here today to obtain a common understanding of A, B and C. I know that D and E happened this morning and that a lot of people here would probably also like to go over F and G, but the plan right now for the next twenty minutes is A, B and C."

  – "I realize that the situation regarding ... could change next week or next month – who knows, it may not change at all – and I realize that this would mean that we would immediately change the product, no question. Today, however, I want to use our time to deal with the present situation, the conditions that we have now, at the moment and not spend time speculating. If things change, then we will need to change, too."

- A rhetoric question, real-life story or example can be a good opening if you want to do a little more.

- Be as authentic as possible. If you are not "happy to be here today", then use another phrase – don't say it.

- Avoid humour if you have a multi-national audience. Test your funny stories with cultural insiders beforehand if you have a single nationality audience. Humour can be very "unfunny" or even offensive if it falls flat. People may feel unsure or even stupid if they don't understand. If you're not sure, don't do it.

- Also avoid apologies ("I didn't really have enough time to prepare the slides") or making yourself appear less significant than you really are ("I'm not really the boss, I just let people do their thing") – these are both typical for the UK but could be seen as "losing face" in a lot of Asian countries.

Introduction components

An introduction needs four components: an opening, objective, overview and some information about organisation.

Openings

- Good morning, I'm ... (*name*), from ... (*department and / or company*).

- Good afternoon everybody. For those who don't already know me, my name's ..., from ...

- Hello everyone. I know most people here already, for those I don't, my name is ... and I'm from ...

- I'm pleased to be here in ... (*place*) with you today, ... (*date*).

- It's good to be here in ... with you on this sunny Friday morning.

- Here we all are in ... this Wednesday afternoon.

- So, let's get started. (*More informal with no name, if you all know each other.*)

> Typical German English mistake: "Good morning altogether" – should be "Good morning everybody" or "Good morning everyone".

Opening with a question

- So, everybody, why are we here? We are here because you need an update on the changes to the booking process. It's important because you will need to use it next week, starting Monday morning at 08.30. My name's ..., from ... and ... (*objective*).

- Good morning everybody. The question now is why are we at Schneider GmbH the best partner for Millwards Systems? The answer is because we really do meet your needs. For those who don't already know me, my name's ..., I'm the technical director at Schneider and ... (*objective*).

- How reliable is our service and maintenance system? The answer is 100 percent over the last six years. That, ladies and gentlemen, is what I'd like to show you today. My name's ..., from ... and ... (*objective*).

Opening with a story

Be as specific as possible, with a couple of dates and times, numbers and background details.

- On the 31st of March this year, we put in our XYZ system at AB Consulting, with over 1,200 users worldwide. There were the minimum of start-up problems and AB Consulting can definitely recommend us and the XYZ system. For those who don't know me, my name's ...

- Hello everyone, I was on site last Wednesday, Thursday and Friday in Kiev. The weather was absolutely freezing, but work went well. There were over 130 workers that week and this week we will have over 160. Things are going well on the whole (*objective*).

Opening with an example or picture

- Next week, we will have a six-figure reference number instead of four figures. This is one of the changes to the invoicing process that we need to look at today. My name's ... from ... and (*objective*).

- This (*illustration on slide*) is my favourite of the new posters in our new marketing campaign (*objective*).

Objectives

- By the end of this presentation, I hope you will ...
- By the end of the next twenty minutes, you should ...
- We now have sixty minutes for you to ...

- ... know why A, B and C are important for you in your job.
- ... have a clear picture of progress on the project including the main challenges.
- ... be able to see why we at Schneider GmbH are the very best partner for you.
- ... have a clear overview of the financial / technical / sales / marketing aspects.
- ... have an update on the changes to the existing system / product / method.
- ... be updated on the latest marketing campaign / sales figures for the third quarter.

When more than one objective:

- ... and, as well, ...
- ... and on top of that ...
- ... and in addition ...
- ... be able to implement this information starting next week.
- ... have enough information to reach a decision regarding supplier / time schedule.

A typical German English mistake is to always say "my presentation" – should be "the presentation" or leave it out. For example, "This is the overview of my presentation – should be "This is the overview".

Overview

- First, we'll look at A, then move on to B and finish with C.
- A is our first point, followed by B and then C is last.
- First, the present status, next, the present challenges and proposed solutions and finally, the next steps.

Organisation

- We have forty minutes in total.
- We have one and a half hours, with a ten-minute break halfway through.
- If you have any questions at any time, do please ask.
- If you have any questions, please could you hold them till the end, when we will have time for a discussion – except questions for understanding, of course.
- Coffee will be served in the break / at the end on the tables outside the doors.
- Matthias has reserved a table at the Hard Rock Café for us all for lunch afterwards.
- Can we all switch our cell phones off? Or at least have them on silent mode?
- If anybody needs to take a phone call, I'd be grateful if they could leave the room.

Many person-oriented cultures, especially the Middle East including Saudi Arabia, Kuwait and others may find it unthinkable to be unreachable during your presentation. Don't even ask them to switch off the phones; if you are important enough they will do it without asking (see "Handling interruptions and disturbances", page 105).

Dealing with handouts

Providing handouts very much depends on the situation and content of your presentation. Remember that a lot of cultures will be very interested in you as a person, as well as the facts and figures. If the handouts are designed to be used during the session then you obviously need to give them out beforehand.

Example

 People from specific countries, such as Japan, will expect handouts in a presentation. Research the Internet when presenting to one or two nationalities.

Handouts at the beginning

- Finally, before we start, there are handouts for everyone to make notes if they wish. Do we need some more? I can see a man at the back waving his hand – could you pass this back to him? Thank you.

- Everybody should, with a bit of luck, have already seen the presentation slides. I sent them out last week. There are some additional hard copies here for anyone who needs one.

- There are handouts for everyone to make notes if they wish.

- There are handouts at the back of the room if anyone would like to take notes.

- Does everybody have a handout? If not, I have some more here.
- Does anybody *not* have a handout?
- Do we need more handouts? If we need them, we can have more copies made.

Handouts at the end

- There are handouts by the door for you to take on your way out.
- If you would like a copy of the presentation, please leave your e-mail address on the list by the door.
- I'll send everybody a copy of the slides for future reference.

Requesting copies

- Could you make us another 20 copies please?
- Could you staple them together with holes? That would be great.
- ... Double-sided would be fine as well.
- ... Hmm, double-sided won't quite work because the print on the diagrammes will show through.

Taking care of technical problems

When something goes wrong you should do what you want your audience to do – stay calm. Use some or all of these steps:

- Keep the audience informed at all times, even if nothing happens.
- Name the problem.
- State the options (depending on the situation).
- Decide on action (any decision is better than none).
- Follow that action including adhering to time limits.
- Stay calm and don't stop communicating!

Examples

"My computer doesn't want to start right now. I can change the power supply and try again or use someone else's laptop with my USB stick. (*Member of audience offers laptop.*) Great. Okay then, I need two or three minutes to set up again. You can have a coffee, make a phone call, etc. ... Okay, let's get started. Thank you for your patience everyone."

"We don't seem to have a connection to the computer. Let me switch off and start again. Can you bear with me? It'll take a minute or two and then we can see if it works ... Okay, I'm afraid it's still not working. Let's take a ten-minute break and meet again at 11.15. I'll try and find another projector in the meantime, otherwise we can do without one ... Unfortunately, we couldn't solve the problem. We can do one of three things now – either postpone the meeting or all crowd round my laptop, or find another room in the next building, maybe. (*Audience wants to crowd round laptop.*) Okay, for those who don't know me ..."

Computer problems

- Well, my computer doesn't want to start.
- It looks like my computer is having problems.
- I'm afraid my computer will need another two or three minutes to restart / boot up.
- My computer is very slow today.

Projector problems

- The projector doesn't seem to be plugged in. Could you check the socket for me?
- We don't seem to have a connection to the computer. Let me start again.
- It looks like there's no signal. Does anyone know how to scan for a signal on this projector?

Sound and light problems

- We don't have any sound. I can't find the command / button – can anyone help?
- Could somebody close the blinds?
- Does anyone know where the switch is?
- Could you close the curtains?
- Can we switch the lights off at the front?
- Can we keep the lights on at the back?

Main section: skills and techniques

You can use a few or all of the clearly-explained process steps in this chapter, together with the lists of useful phrases to improve your presentation delivery, question handling and disturbance management as follows:

- fixing your body language and using your voice well in the foreign language (page 58 and 62),
- making transitions (page 65),
- explaining slides and diagrammes (page 70),
- dealing with questions (page 87),
- handling interruptions and disturbances (page 105).

Fixing your body language

International viewpoint

When presenting to different nationality audiences you need to be aware of three main body language points: eye contact, smiling and the audience's body language.

Example

> When presenting to the Japanese, make sure you know who the most senior person is and address your presentation to them. Japanese respect for hierarchy is usually very high. Do not try to motivate others to ask or answer questions; this could be embarrassing for them. Discussion, if any, should take place with the senior manager or, most often, after the presentation out of work hours, during perhaps karaoke evenings.

Eye contact

If the people you are speaking to don't look you directly in the eye, don't force them to do it by trying to move into their view. Direct eye contact in Western European countries most often means honesty and friendliness, but not always in other parts of the world. In Japan and other Asian countries it can mean a lack of respect. Men and women looking at each other directly in the eye can also mean different things in countries such as Turkey, for example.

Smiling

Smiling at the beginning and during your presentation is seen differently in different countries. Americans are likely to smile more than Germans, who are likely to smile more than

Russians. Indeed, in Russia a smile is not a sign of politeness, nor is it common to smile when meeting a person for the first time, or at the beginning of a presentation. A Russian saying goes: "To smile without reason is the mark of an idiot." Important to know, too, is that Chinese and a lot of Asian cultures will smile or laugh if embarrassed. Inform yourself beforehand for presentations to one nationality; otherwise use your smile sparingly. Try mirroring others – smile if they smile at you.

Audience's body language

Don't be put off if your audience doesn't react the way you expect. Indeed, German audiences can be seen as non-smiling immobile or motionless "stones" by UK/US presenters. Indian audiences may "waggle" their heads – from a western European point of view, as if they are disagreeing, but they are in fact probably agreeing. The UK/US and others may nod their heads from time to time, smile, throw in "agreement" comments or noises – if they sit completely still and unsmiling, this could mean disagreement or even hostility. Some Japanese may look as if they are falling asleep – it could be that the presentation itself does not have the same value as in Western Europe, more important could be the talks later, outside the presentation room.

Be careful of differences in body language. You can watch the TV, if presenting in that country or research for "taboos" from books, the Internet or a cultural insider. Don't sit with the soles of your feet showing in the Arabic world, don't point directly at people in Britain, the finger and thumb "okay" sign in Germany means you are going to kill someone in Tunisia, to name but a few.

Body language basics

When you make international presentations, it really is very difficult to know all of the culture-specific body language rules of your audience. Even if you do, you then have the dilemma of a mixed nationality audience – which rules should you follow? The answer is to stay as natural and as neutral as possible. The following tips are the most important.

Checklist: body language basics

Preparation	When preparing for a presentation, stand up and say it at least once. No time? Then, at the very least, the words of the beginning, introduction and the summary should come out of your mouth at least once before you do the real presentation. This will help you feel and therefore look much more confident.
Clothing	If unsure, go overdressed. Again, watch for the way things are done. In the UK and US, no tie, taking a jacket off and rolling up sleeves might be okay, but Japan and Saudi Arabia usually expect suits and ties.
Hands	Do hold your hands at waist height or above to start and finish your presentation – waist and above looks strong, below waist looks weak. Again, watch for the way things are done – e.g. Belgians and French may move their arms from

	the shoulders upwards and not just from the elbows. Be aware and choose what you do.
Eyes	Don't look at the slides and the walls, do look at the people (or in their direction, if they avoid eye contact). Don't focus on one person you like or who looks most interested – you might appear to be excluding the others. Do look at the audience zig-zag style for a couple of sentences at a time.
Feet	Don't stand with your feet tightly together – this looks unnatural and stiff and can be interpreted as nervousness about the subject. Do keep your feet a little apart; one can be further forward than the other – then you look balanced and solid. Don't point your feet away from the audience – this can be seen as not wanting to be there, or even not liking the people. Do point your feet towards the audience in an open "v", like clock hands at five to one (12.55) – this usually looks natural and friendly.
Move-ment	Don't stand still, rooted to the same spot the whole time – it's unnatural. But don't dance around constantly – it can be tiring for the audience to concentrate on your words if you move all the time. Do make definite movements, from one specific place to another, then stop and talk. Move again, stop and talk again.

Using your voice well

International viewpoint

Speak more slowly than usual and watch your audience's reactions, if you are presenting to an audience of non-native English speakers. Make breaks between phrases. This is easier than speaking more slowly over a length of time. Watch the audience carefully and repeat yourself using other words, if necessary. Watch for differences in volume – Greeks, for example, tend to be louder than Scandinavians – and switch style if you choose.

Two key techniques

We are sure you must have seen sentences that go on and on without any full stops or commas or capital letters and are so long that they are difficult to understand and sometimes even go over three lines or more and use a lot of difficult long words and phrases that are not needed and just go on and on and on and on and on ... Do you see what we mean?

Basically, two techniques are enough for you to be sure your voice sounds interesting in English. If your spoken English is at elementary level and not intermediate or advanced, using these two techniques can make you sound much better.

Use pauses

Firstly, use pauses. Pauses are quite wonderful for a lot of different reasons:

- they give you time to think,

- they give your audience time to understand (a lot of them are non-native speakers, too),

- they can create a little or a lot of suspense and drama (interesting),

- they can replace any "ums" or "ahs" you might want to say.

Stress key words

Secondly, put stress on some of the key words. That's like giving the music of your text some rhythm – a bit of a beat.

Exercise

 Stress the marked words and hear the difference – the above text could "look / sound" like this:

We are **sure** you must have seen sentences... that go **on** and **on** ... without any **full stops** or **commas** or **capital letters** ... and are so **long** that they are **difficult** to **understand** ... and **sometimes** even go over **three lines** or more ... and use a **lot** of difficult long **words** and **phrases** that are **not needed** ... and just go **on** and **on** and **on** and **on** and **on** ...

Advanced techniques

If you wish, you can make your presentation more interesting and have some fun with your voice at the same time. You don't need to be an actor nor sound like someone different. Stay yourself, but try "stretching" yourself, to gain more impact and get your message across to your international audience even better.

Put contrast in your voice

The problem is that same speed, same volume and same tone can be monotonous. A monotonous voice is usually boring and difficult to follow. Contrast is the key. It's not difficult. You can put contrast into your voice in these three areas:

- speed (fast vs. slow)
- volume (loud vs. quiet)
- pitch (deep vs. high)

Exercise

Try it out on this text for fun (or on your own presentation). Read the first line fast, the second slowly, the third fast, etc. Then read it again, the first line loudly, the second quietly, the third loudly, etc. And finally, again, the first line in a deep voice, the second high, the third deep, etc.:
We are sure you must have seen
sentences that go on and on
without any full stops or commas or capital letters
and are so long that they are difficult to understand
and sometimes even go over three lines or more
and use a lot of difficult long words and phrases that are not needed
and just go on and on and on and on and on...

A final point – if you find your time is limited and you have only 15 minutes to do your 45-minute presentation, talking faster won't work. You will not finish on time because you probably can't speak English as quickly as German. In addition, the audience will find it very difficult to follow, especially if they are non-native speakers. Decide quickly what percentage of 15 minutes each part of your presentation should take. Keep your eye on your watch and limit yourself to the key concept in each portion. Next time, be prepared. Think about what you'll keep in the presentation if your time is divided in half or if you are only given 5 minutes.

Making transitions

German speakers often use long, formal phrases to make links. You don't usually need them when presenting to an international audience. Fewer words often make more impact.

Between slides

Rhetorical questions are a simple method of linking, but don't overuse them. Alternatively, state the content of the slide with little or no introduction.

Useful phrases

- What's next? (*click to next slide*) The technical features.
- The technical features are next.
- The technical features are shown next.
- Next, we have the technical features.
- These are the technical features.
- Here are the technical features.
- The next slide shows the technical features.
- What options do we have? This is the first of three possibilities. (*click*)

- Why is A important? This is why. (*click*) What is the current A? Here it is. (*click*) What does the future A look like? This is it. (*click*)

- What is the current status? Here it is. (*click*) What are the figures? Here they are.

- Where are we now? Here we are. (*click*) What do the figures look like? Here they are. (*click*)

- What are the benefits? You can see them here. (*click*)

- What is the technical functionality? You can read the description here. (*click*)

- How did we solve the problem? This is how. (*click*)

Between sections

Making clear links between the introduction and main section and between sub-sections can help your international audience to follow your presentation more easily.

Useful phrases

- Let's get started.

- Are we ready to start? (*introduction*)

- Good. Next comes the ...

- Okay. So ...

- So. Our first point – the worldwide total sales last month ... So far, so good. Our second point is ...

- Let's go to the next section, the time schedule for the data migration.

- Let's move on to the service level agreement.
- Moving on to the qualifications needed ...
- Now, we'll look at the summer campaign.
- I'd like now to turn to the year-on-year analysis.
- Next, I'd like to look at the process steps.
- The new range of summer cosmetics is next.
- (*Show agenda slide*) Where are we? We have covered A and B in the last twenty minutes, now I'd like to look at C before the break.
- So, we have looked at A and B, now let's go on with C.
- After Europe, the Middle East and Africa, let's move on to the Americas.
- Now we have a detailed picture of the problem, let's look at the options on offer.
- We looked at the content of the migration and the suggested time schedule. Let's go on to look at the manpower and distribution of tasks.
- Okay. We outlined the hardware features and showed you the network system. Let's go on to service and maintenance now.
- Good. That was the old system and then the new system. Now, let's look at the advantages of the new system one-by-one.

Between breaks

Breaks of ten minutes or less are usually impossible for people to come back on time. Try fifteen or twenty minutes and allow for an extra five minutes on top to be sure. If you announce a clear time to continue, your chances of having everyone back are better.

Announcing breaks

- (*Show agenda slide*) We have covered A, B and C. Now let's have a break before we go on to D and E.
- That was the first half, now it's time for a break before the second half.
- Okay. We looked at the current figures for all 46 product lines and analysed them compared to last year's figures; now, I think we all need a coffee / some refreshment.
- Good. That's all for now. A and B are covered; we'll look at C and D after lunch.
- Can we all be back here in twenty minutes? That's eleven-twenty by my watch.
- Let's start again at two-forty-five.
- I'd like to carry on at three-fifteen.

Giving information about refreshments, lunch, etc.

- Coffee, tea and cookies are at the back of the room.
- Refreshments are just outside the door.
- You can order a drink from the waiter / waitress.

- We can have a drink at the bar.

- Lunch is in the canteen in the next building. Please don't forget to take your visitor pass with you.

- For those (people) here for the first time, the canteen is on the second floor in the next building. Just follow the others.

- Lunch is booked in a Greek restaurant in the Schellingstraße. I suggest we form small groups to walk over there together.

- Who knows the way to the restaurant? Okay then, for those who don't know the way, just follow one of these people.

After the break

If people are missing, you can either start, wait a while or ask your contact or audience what to do. It depends on the situation. In China, France, Italy, Japan and Saudi Arabia, among others, it is usually necessary to wait for the most important person. This is, of course, also true in other countries. It normally doesn't make sense to start without the main decision maker. It is always important for you to keep your audience informed, so that everyone knows what is happening.

Useful phrases

- Okay, let's carry on with point C.

- Good, let's continue.

- I'd like to go on to the final section, now.

- So, what's next?

- It's now eleven-twenty. Not everybody is back yet, but I'd like to carry on anyway.

- Okay then, it's eleven-twenty and there are a few people still missing. Let's give them another five minutes and start again at eleven-twenty-five.

- It's eleven-twenty. Would you like to continue or wait for the others? I don't mind.

> US English "I don't care" means the same as UK English "I don't mind". UK English "I don't care" is very negative – use "I don't mind" or "It's all the same to me" for international audiences.

Explaining slides and diagrammes

Don't compete with the slide. Your audience will always try to read whatever is in front of them, regardless of what you are saying. The next four steps can help you get your message across effectively – especially when your audience includes non-native speakers:

1 Announce your slide.

2 Show your slide.

3 Let people read it, before you start speaking. During this time, you should demonstrate what you want the audience to do – you should turn to your slide and read it silently yourself.

4 Speak about the most important point(s) of the slide and / or give details, examples and real-life stories about those main points.

Text slides

Be sure to state the main point first, before going into specific details – you can think of this as "big picture versus little picture".

Examples

What's important here? The main point is that the figures are positive on the whole. We went over 36 of 54 product targets; we were just under 12 targets and missed 8 targets by between 5 and 13%. However, the total is positive.

Data availability is the most important of these six key strategy points for us here today. It's fundamental. If we don't have the data, then we can't negotiate better prices. For example, it's far easier for us to get better rates with an order volume worth € 300,000 than with an order volume worth € 30,000. It makes sense.

The main point here is that we are represented in 12 different countries worldwide.

We have our headquarters in Austria, factories in the Czech Republic and Singapore, as well as branch offices in another 9 countries throughout Europe, the Americas and Asia.

Do not read the slides word for word. This can be very dry and boring for your audience. Let them read themselves. You choose the main or most interesting point(s) and give details, an example or real-life story about only those point(s).

Useful phrases

- What's important here? The most important thing is ...
- Why is this interesting to us? Well, it's interesting because ...
- What's unusual here? It's unusual because ...
- XYZ is the most important message here. Let me tell you why ...
- XYZ is the most important point. When consumers make the decision to buy ...
- The most important message is ...
- The main point here is ...
- The main difference between this system and the old one is ...
- We have 12 points on this list – the most important two are ...

Charts and diagrammes

A graph has an x and a y axis with lines or curves. If it has blocks then it's a graph or a chart. A diagramme is a more general word and can often be used for different types of graphs and charts. You don't need to know the name of your chart, diagramme, graph or table. You can just say "Take a look at this" or "Have a look at this", followed by "As you can see, here ... and here ..." US prefer "take", UK prefer "have".

As before, be sure to explain the "big picture" first, before going to the "little pictures":

1 Main point
2 Explanation of layout
3 Highlight key points

Examples

Here are the total estimated savings in thousands of euros for this financial year. The black section shows savings already achieved and the grey section shows savings to be achieved by the end of the financial year.

Here we have sales by product line (*point to x axis*) in hundred thousand euros (*point to y axis*). The black line shows last year's sales for each product; the blue line shows current sales and the red line shows the target. So ... (*give your audience time to look*) ... we have 20 product lines here – two are not going very well, six are going alright, ten are going fine and two are going very well indeed. Let's look at the first two and the last two in more detail ...

Main point

- Here are the total sales shown in hundred thousand euros.
- This is what the XYZ should look like in the future.
- This is a summary of the XYZ data.
- This is all of the ABC put together.

Explanation of layout

- Up here / down here we have the ...
- On the right is / are the ...
- On the left is / are the ...

- In the middle is / are the ...
- Top right / middle right / bottom right
- Top left / middle left / bottom left
- This axis shows ABC along here (*point*) against DEF up here (*point*).
- This graph shows total sales, with ... up on the left and ... across the bottom.
- This chart shows FY2006 compared to FY2007, with ... down on the left and ... across the top.

Highlighting key points

- The three problem areas are highlighted in yellow.
- I highlighted the three problem areas.
- The top performer in each section is in bold.
- See the next chapter on how to interpret and explain connections between the data.

The German word "Kurve" can mean a curve (the line showing the results) or a graph (the diagramme with an x and y axis). Be sure you use "graph" to talk about the whole diagramme and "curve" to talk about the line showing the results.

Business English terms

Business English is definitely not the same worldwide. Differences exist between American, British, Australian, South African English and many, many more. What does this mean for you when presenting internationally? You shouldn't learn any complicated expressions or local sayings – you risk making unnecessary mistakes or, even worse, saying the expression in the wrong tone of voice, and losing the meaning completely. Simple, well-used words are best.

Reasons behind events

Examples

> The increase in revenue is a result of a great many factors. The main ones are first the products themselves - the current range is excellent, next is the fact that one of our competitors Company ABC went bankrupt and the final major factor is that the exchange rate is currently very favourable for exports.

> The decrease in profitability is due to a number of different reasons. First, we had a raw material price increase of 5% in October, secondly, we opened the new warehouse in November and finally, we had to recall the XYZ product at the end of the year.

> The rise in costs is mainly due to the raw material price increase.

> The increase in amount of services on offer is because of the customer need for flexibility.

> The product range diversification is partly a result of the customer survey findings.

> The drop in revenue is largely due to the fact that the weather was so bad.

The fall in the number of new customers is because we have new competition.

Useful phrases

- This is due to + *noun*
- This is because of + *noun*
- This is a result of + *noun*
- This is due to the fact that + *sentence*
- This is because + *sentence*
- This is the result of the fact that + *sentence*

Useful vocabulary

current	aktuell / gegenwärtig
because of	infolge / wegen
due to	ist auf ... zurückzuführen / wegen
a result of	ist ein Ergebnis von / der / des ... / wegen
largely	im Wesentlichen / zum größten Teil
mainly	hauptsächlich / in erster Linie
partly	teilweise / zum Teil

Results

Examples

So, we know what happened, and why – the question now is what next?

All of this means we will need to work even harder next quarter, to make sure we keep our existing customers.

As a result, we will need to revise the next quarter forecast.

This could lead to the need for a new system.

This may mean cutting down on the number of products next year.

Useful phrases

- All of this means ...
- As a result we will ...
- This could lead to ...
- This may mean ...
- This may result in ...
- ... will be a direct result.

Change and development

Examples

What major changes were there compared to the last quarter? As you can see, there was a huge increase in sales, due to the good summer weather. We are 25% up on the previous three months and 5% over target, which is very good news indeed.

What are the most recent developments? Well, there are two points I'd like to highlight – first, ... and second, ...

What main differences can we see? From the user point of view, none really. From the IT point of view, the interfaces will change, with the Global Framework Platform becoming the single central data exchange point.

The number of on-time deliveries increased a little.

The figures improved at the start of the year.

The number of new customers decreased steadily.

The amount of claims rose substantially.

Return on investment (ROI) is not as high as last year.

Shareholder value is not as low as this time last year. In fact, it's 3% up.

There was a slight drop in the number of existing customers.

Profits fell slightly in the plastics division.

The quality level stayed steady.

Despite a great deal of fluctuation, the overall result was positive.

The figures fluctuated a lot, but the situation is stable now.

"What" is used when you have a non-limited choice – "what sort of food do you like?"; "Which" is used when the choice is limited – "which sort of food do you prefer, Italian or Chinese?"

Useful phrases

- There was a small increase over the last three months.

- ... is 25% up on last quarter.

- I'd like to highlight three main points: first ..., second ... and third ...

- There was a slight drop in the number of ... / the amount of time spent ... (-*ing*).

- The ... will change and the ... will become ...

- The ... is likely to cause the most challenges.
- The ... increased / rose a little / a lot.
- The ... improved at the start / in the middle / at the end of the year.
- The ... decreased / fell steadily / slightly / substantially.
- The ... is not as high / low as last year.
- The ... stayed steady compared to last quarter.
- Despite / in spite of a / the ..., the overall result was positive / neutral / negative.
- The ... fluctuated a lot, but the ... is stable now.
- The ... is quite unstable.

Useful vocabulary

a decrease / a drop / a fall	Abnahme / Rückgang
fluctuation / to fluctuate	Schwankung / schwanken
large / substantial / a lot	reichlich / eine Menge
to raise	anheben / erhöhen
a rise	Steigerung / Anstieg
to rise	ansteigen
small / slight / a little	klein / geringfügig / ein wenig
to stay steady to remain constant	gleich bleibend / unverändert konstant / beständig / gleich bleiben

Please note:

- The words "rise" and "raise" often cause confusion: the noun "a rise" is used for a price rise, a rise in productivity, a sunrise and can also mean an increase in salary. The noun "a raise" usually means an increase in salary – "Gehaltserhöhung".

- The irregular verb "to rise – rose – risen" is used to express that something happened: prices rose by 8%, productivity is rising. The regular verb "to raise – raised – raised" is used to say that something or someone made something happen: we raised the prices by 8% (**not**: we rose the prices by 8%); productivity was raised by cutting the number of process steps (**not**: productivity was risen by cutting the number of process steps).

- In German, "die Fluktuation" specifically means employee movement whereas "fluctuation" in English means any type of movement, more like "die Schwankung".

- Adjectives describe a noun – a slight increase, a substantial drop. Adverbs describe a verb – the figures increased slightly; overhead costs dropped substantially. "-ly" is often added on to the end of an adjective to form the adverb. German speakers mostly add "-ly" too often – if you are not sure, leave it out.

Problems

Problems and bad news are often difficult to present, not only because first, you need to keep an international viewpoint in mind, and second, you need to think of the words of the English language, but also because of existing internal and external company politics and relationships that you are dealing with. Choose from the following examples and phrases, depending on your specific presentation situation.

International viewpoint

Use the cultural checklist on page 28 to decide on your strategy. Be careful. Talking openly about problems or admitting bad news could mean losing face in a lot of Asian countries. A presentation might not be the right thing – ask your cultural insiders – you might have to discuss bad news one-to-one and not in front of a group. Once you decide on your tactics, don't forget that a lot of the rest of the world is likely to be more indirect than a "typical" German speaker. However, although both British and North American speakers value politeness, North Americans are generally more straightforward and can appear impolite to the British!

English language points

When presenting problems, it's quite usual not to use the word "problem" at all, except perhaps in a positive sentence ("we fixed most of the problems"). Indeed, talking about "problems" can be seen as very negative, or even extremely direct and harsh from an international point of view. Look at the examples for different options.

- Use negative opposites such as: not good (*instead of* bad), not smooth (*instead of* rough), not simple (*instead of* complicated), not reach (*instead of* miss), not easy (*instead of* difficult).

- Use softeners such as: actually, in fact, well
(at the beginning of your sentences, especially when answering questions. This will make you sound less abrupt).

- Use qualifiers such as: a few, a couple ..., a number of ..., quite a few, a bit, a little, very... (to sound more rounded and less staccato).

Examples

Actually, we had a few minor setbacks at the beginning, nothing unpredictable, but a lot of people needed to do a lot of work in a short space of time to make sure that the system stayed up and running. The situation wasn't good for a number of days but then stabilised.

In fact, there were a number of challenges on the way which caused quite a few headaches. These included a lack of information, insufficient communication and unclear allocation of responsibilities.

Well, not everything went smoothly, but we were able to sort out most of the points in the first couple of days. We didn't have any major breakdowns, but a couple of short machine stoppages.

Some elements were not as simple as foreseen and needed a lot more manpower than planned.

Despite very thorough planning, we underestimated a bit the time needed for the commissioning phase. We fixed this by bringing in a second team of engineers, but had to bear the costs ourselves.

In spite of extremely careful planning, we overestimated the amount of materials needed and ended up with a surplus.

The target wasn't quite reached this month, but I'm sure we'll do better next month; the figures are looking much better this week already.

Useful phrases

- Actually ... / In fact ... / Well ...
- We had / There were a few minor setbacks.
- We had / There were a number of challenges.
- We didn't have / There weren't any major breakdowns.
- A lack of information
- Insufficient communication
- Unclear allocation of responsibilities
- ... to make sure that the system stayed up and running / didn't break down.
- ... we could fix / sort out most of the points.
- This was more difficult / complicated than foreseen.
- It was more expensive / time-consuming than planned.
- We underestimated the number of ... (*countable items*).
- We overestimated the amount of ... (*uncountable items*).

Useful vocabulary

allocation	Aufteilung / Verteilung
to bear costs	Kosten tragen
challenge	Herausforderung
to fix things (problems)	etwas in Ordnung bringen

to go smoothly	reibungslos über die Bühne gehen
to overestimate	überschätzen
setback	Rückschlag
to sort out problems	Probleme lösen
surplus	Überschuss
thorough	gründlich
to underestimate	unterschätzen
unpredictable	unvorhersehbar / unkalkulierbar

Making comparisons

When making comparisons in English, make sure you know the key words you need.

Examples

The first option is more economical than the second because it needs less maintenance. However, the initial costs would be higher.

Splitting the delivery would be uneconomical and relatively complicated. We think the best thing to do is to delay the total shipment.

This new development is extremely useful. It cuts the number of process steps by half, compared to the previous method. It's much simpler than before, as well as being more accurate, faster and more reliable.

Option A would be extremely time-consuming. Option B is the quickest, most effective and straightforward method. Even if it costs more, it will save money in the long term.

Useful phrases

- The first option is more economical than ...
- However, the initial costs would be higher.
- ... would be uneconomical ...
- We think / It seems the best thing to do is to ...
- This is more accurate / faster / more reliable compared to the previous method / system / product.
- Even if it seems to cost more / need longer planning / need more maintenance, it will save money / time / manpower in the long term / short term.

Useful vocabulary

to delay	(sich) verschieben
economical	wirtschaftlich
long-term	langfristig / längerfristig / auf lange Sicht
maintenance	Wartung / Instandhaltung
reliable / unreliable	verlässlich / unzuverlässig
short-term	kurzfristig / kurzzeitig
time-consuming	zeitintensiv / aufwändig
uneconomical	unwirtschaftlich

Plans and goals

Examples

 This slide shows current progress, with the tasks listed on the left and the time along the bottom. As you can see, most of the building work is on schedule. The electricity lines and water pipes are actually a few days ahead, but some of the piling is two weeks behind schedule. The drawings were delayed and we had to postpone some of the work. We hope to catch up by the end of August.

Overall we expect to hit the total revenue target by the end of the financial year. We missed the target in the first quarter by just 20,000 euros; we're below target again in the second quarter by 30,000 euros, but recovered in the third quarter and exceeded the target by more than 60,000 euros, which put us back on course for the end of the year.

Useful phrases

- The ... is ahead of / on / behind schedule.
- ... is a few days / weeks / months ahead / behind.
- The ... were delayed.
- The ... was postponed.
- We should catch up by the end of the week.
- The plan is to be back on course / target / schedule by the end of the month / year.
- We expect to hit target.
- Everything is on target.
- The ... target was missed by only 15 units.
- We were below the ... target by just 15 units.
- The ... recovered in the third quarter.

You can pronounce "schedule"as "SKedule" or "SHedule" – both are correct.

Useful Vocabulary

ahead of / on / behind schedule	dem Zeitplan voraus / im Zeitplan / im Verzug
to bring forward	*hier:* vorverlegen
to cancel	absagen
to catch up (time)	Zeit aufholen
to delay	(sich) verschieben
to hit a target	ein Ziel erreichen
to miss a target	ein Ziel verfehlen
over / on / below target	über / im / unter Soll
to postpone	verschieben / zurückstellen
to recover	*hier:* einholen

Dealing with questions

Handling questions well requires a lot of different skills. As with presenting problems and bad news, a great deal depends on your situation and the balance of power – especially the question of whether to take questions throughout your presentation (mostly recommended) or save them until the end.

International viewpoint

The interaction between you and your audience could be very different and surprising to you. This is because of the following four main points:

- presenter vs. audience responsibility – who gives or obtains the information?
- structure – should questions be taken anytime or at the end?
- timing – do you need longer to allow for the foreign language?
- speech patterns (turn taking) – how can you interact best with international listeners?

Presenter vs. audience responsibility

Your responsibility as presenter changes in different parts of the world. In Germany the presenter is responsible for the audience receiving and understanding the information presented. This often means a lot of time, first, thinking about what exactly to present and second, preparing slides. These carefully prepared slides are often, in the presenter's opinion, never good enough. It also most often leads to a great deal of detail, background information and general content – slides packed full of writing – usually to ensure that nothing is left out. A lot of questions could mean that you didn't present the right information.

In other parts of the world, America or Britain, for example, it's the other way round. The audience is responsible for obtaining the information. It's their job to get what they want. This may mean that an American presenter will prepare a number of slides, but not worry too much about details because they will go with the flow – depending on what the audience needs, depending on which questions they ask – and will usually be quite happy to show slides from other presentations.

Your French, Italian and Spanish audiences may find your amount of content detail very "dry" and "uninteresting", preferring to ask questions and have a discussion based on key points in slides. On the other hand, of course, presenters from other countries may appear "unprepared" to you – with little detail in their slides and not enough background information or analysis.

Examples

When you present to North Americans or Europeans in general, don't be surprised by a lot of questions. This usually means a high degree of interest.

Presenting to a group of Japanese can be very different. The more important the people you are presenting to, the more people they are likely to have with them. They may be surprised if you are alone and wonder where the others are. It is not unusual for the audience to close their eyes while listening. You may not be asked any questions at all. Do not try to ask specific people if they understand or if they have any questions, this could be much too direct and very embarrassing, especially if they are not the most important people. It might be out of place for them to speak. The discussions and decision-making are unlikely to take place during the presentation, but more likely afterwards, between you and the most important people (as long as you are on the same hierarchy level as they are), possibly at a restaurant or in a bar in the evening.

Structure

If you request your audience to ask questions only at the end of the presentation, this could be very frustrating to poly-chronic cultures. When presenting to international audiences it is better to allow questions anytime. See page 12 for more information on polychronic and monochronic cultural groups or page 33 regarding circular and linear structure.

Timing

Whether or not you choose to make a presentation with a lot of detail, you most likely need more time for questions than when presenting to fellow nationals. A good rule could be to plan 50% of the time for presenting and 50% of the time for questions. Longer presentations of one hour or more can be planned with the 50% rule for questions anytime, but with a back-up or additional section such as a case study, if needed. If you plan like this, the worst possible thing that could happen, is that you finish sooner than expected – this would probably be welcomed by your audience and not be a bad thing at all.

There are a number of possible reasons for needing more time, including:

- Some cultures such as French and Italian simply prefer discussion and talking together more than a presentation in which only one person speaks at a time,

- Many cultures, such as those from the Middle East, see time differently. There is always more of it and there is not the same need to keep one's comments short and to the point, as desired in Germany.

Speech patterns or "turn taking"

When taking questions, you need to be aware of speech patterns, also called turn taking.

The standard **German speech pattern** (in general, as always of course!) is as follows: person A speaks. When they finish, person B speaks. When they finish, A speaks again. There are usually no pauses between the speakers and they are not expected to overlap each other. Overlapping could be seen as impolite.

A			
B			

The standard **North American and British speech pattern** is as follows: person A speaks. When person B thinks they are coming to the end of their sentence because the tone of the voice changes and A makes eye contact with B, B starts to talk. When A thinks they are coming to the end of their sentence, A starts to talk. There are usually no pauses between the speakers and they overlap each other a little. A little overlap is expected.

A			
B			

The **standard polychronic, mostly southern European (inc. French) and Middle Eastern speech pattern** is as follows: person A speaks. Person B agrees with them and speaks, too. A may make a pause and then continue speaking. C talks, too, continuing the same topic of conversation. There are usually no pauses between the speakers and they often over-lap each other. Overlapping is expected. Indeed, no overlap-ping could be interpreted as lack of interest and enthusiasm, or understanding.

A				
B				
C				

The **standard Japanese (and also a lot of Asian cultures) speech pattern** is as follows: person A speaks. When they finish, person B thinks about the words that were said, thinks about the words they want to say and then speaks. When they finish, A thinks about the words that were said, thinks about the words they want to say and then speaks. There are usually pauses between the speakers and they are definitely not expected to overlap each other. The pauses show respect, first, for the importance of the previous speaker's words and second, for the importance of thinking carefully before speaking.

| A | | | |
| B | | | |

So, when you take questions, be aware of the speech patterns which influence when you should start your answer. With the first and the last speech patterns, you should normally wait for the speaker to come to the end of their question. With the second speech pattern, you should also wait for the speaker to finish, unless they continue at length.

If you find they speak at length, then you can use the technique for polychronic speakers. If you try to wait for polychronic speakers to come to the end of their question, you might wait a long time. Pick up on their words and repeat them, then commence to answer.

Example

A: Do you think you could tell me exactly why we are doing this? What is the point of this change? The last new system was only introduced two months ago and we are still learning it. I mean, we always have new systems and new methods, (*) it's never-ending. I think the company just doesn't think about the impact that this change could have....

B: (*starts speaking around * slowly and softly at first, then normally*)... new systems and new methods, yes ... yes... well, the fact is that we need to stay ahead. None of us really likes change, especially the older we get, the more we like to stay in our routines. It's human nature. We need to change to keep up-to-date, to stay ahead.

If you are hoping for questions from the Japanese, you need to say nothing and wait for a while. If there is a longer silence, then you can continue.

Checklist: questions / international viewpoint

You should realise that taking questions at the end of the presentation may be your preferred way (if not, no problem!), but not necessarily that of others. You should

- not be surprised if there are lots of questions from North Americans and Europeans,

- not be surprised if there aren't any from Asians – the business will be done later,

- usually offer to take questions anytime,

- be prepared to answer questions anytime,

- allow a lot of time for questions, 50% for presentations less than one hour,

- be aware and react accordingly to speech patterns when taking questions.

English language points

If you cannot understand the questioner, ask again, ask the audience for help or suggest that you meet later to discuss the question together. Don't forget – always make sure that you assume the problem: "I don't understand." or "That was too fast for me." (see page 22 for more details).

> In German, you usually say "bitte?" when asking someone to ask their question – directly translated into English as "Please?" is wrong! The word in English is usually "Yes?" often with an open hand raised in the speaker's direction. "Yes?" is usually sufficient, perhaps with the name of the person for small groups if you know them.

You may require some of the following phrases for larger audiences. Phrases for opening a discussion are to be found in "Ending components" on page 116.

Useful phrases

- The gentleman / lady at the back / in the middle / at the front.
- The gentleman / lady on the right / in the middle / on the left.
- I can see someone with a question at the back.

Delaying questions

Whenever you can, give a short answer to a question. If you feel you really must ask the speaker to wait until later, be very careful that your tone of voice is friendly and that you don't sound like a schoolteacher. "Please can you wait until later?" is technically polite, but often sounds wrong with a group of adults in business. Make an "I-statement" (Ich-Botschaft) whenever possible.

Useful phrases

- Can you tell me why ...?
- I'd really like to cover that later and finish the analysis first, because it's quite complicated.

- Yes, but just briefly, why ...?

- Well, I don't have a short answer I'm afraid. If I can just go on for another few minutes, everything should become clear.

- I want to be sure you have the full picture.

- I'd like to be sure that we all have the necessary information.

A lot of difficult questions or disturbances can be avoided by
1. not fighting and
2. concentrating on the "here and now" with
3. I-statements as necessary – "Yes, D, E, F are important, I agree, but right now, in this meeting, I want to concentrate on A, B, C."

Preparation and procedure

You need to do just these three things – they really work:

1 Question preparation	Briefly write down all of the most difficult, unwanted questions you can think of and then practise the answers.
2 Five-step technique	Use the technique in the next section when answering.
3 Attitude	Stay calm. Don't fight. Not ever.

1 Question preparation

Briefly write down all of the most difficult, unwanted questions you can think of, each one on a separate card, then pick out cards at random and practise the answers. At the very least, write a list of keywords of difficult, unwanted questions and think through the best answers. Don't just hope that the questions won't come (that can be very stressful). Make sure you know what you will say if they do come.

Types of questions with possible answering strategies:

- **Nice questions** – these are questions that you would like to answer. Use the technique in the next section and be sure to answer clearly and to the point: "You're asking about the number of defective units produced per month – we have between 5 and 10 returns per month, which we replace with a new unit."

- **"Don't know" questions** – these are questions that are relevant, but you don't know the answers offhand. Look around to your colleagues to see if they know; if not, offer to provide the information later. Don't just say "I can find out and tell you later"; you need to be specific to be credible: "You're asking about month-on-month figures for sector C. I don't have the figures with me, but if you would like to give me your business card at the end, I can send them to you next Monday when I'm back in the office."

- **Hypothetical questions** – these are questions about what would happen if … if … if … This is something you don't know, or cannot really say. You have two options. Option one: you say you don't know and focus on the present: "What would we do if ABC happened? I don't know – we would change our plans, for sure. At the moment we're concentrating on fulfilling the requirements of the plan we have now."

Option two: give an answer, but make it clear that it's just speculation: "What would we do if ABC happened? *If* that happened, then maybe we *would* start to produce a matrix machine, but that's just speculation on my part. It would require a whole decision-making process involving a lot of people. It wouldn't be just my decision."

- **Multiple questions** – These are two or more questions from one person at the same time. Your reaction depends on who's asking. If it's someone very important, ask them to talk slowly while you take notes, then answer them one by one (it's extremely difficult to remember them off by heart). If you think the person is less important, then you can

  a. choose the question you would like to answer,
  b. answer the first question or
  c. answer the last question.

  If they don't forget the other questions, they will ask again. The responsibility lies with them to remember the other questions.

- **Internal political questions** – these are questions asked for a specific internal political reason. Pause and think before choosing to answer or not. You can

  a. answer the question carefully and as neutrally as possible,
  b. say that those decisions are not made by you (if it's true) or you're not the right person to talk about that

or it's not the right time and place: "Yes, I know what you mean and we can talk about it together again over lunch today, if you want, but that's the way it is at the moment and I really don't want to discuss it now. I'd like to concentrate on showing you the next steps for putting in the final three systems."

c. just say you'd prefer not to talk about it and then come back to the present: "Well, I don't think I'm the right person to talk to about that, I don't make those decisions. What I *can* say is that we now have the system in place in 12 out of 15 locations, which is the job we were asked to do."

- **Irrelevant questions** – these are questions that don't seem connected to your subject. Be sure to repeat the question – maybe you didn't understand it correctly. Then answer it politely and briefly if you can. If you can't, refer them to the person or department that can, or ask if you can talk about it at a later time together: "You'd like to know about ABC? Could we maybe discuss that together at the end because I'd like to concentrate on DEF for now." (Your presentation is about DEF, you work on ABC, too, but that's another project.)

- **"Stupid" questions** – these are questions that seem stupid or pointless. Answer them concisely and move on. Again, never make your audience seem unimportant. The questioner has some reason for asking, even if you don't know what it is. You could ask after answering, if it is suitable or useful: "You're asking about switching off the machine. Yes, it really is off when it's switched off. The power supply is cut. Do you have a particular reason for asking?" Be careful your tone of voice is not sarcastic; this could be the person responsible for fire and safety, who is very interested in such details.

- **Confidential questions** – these are questions about confidential information. Explain politely but firmly that you cannot answer that question: "I'm afraid I can't answer that; it's confidential." or "I'm sorry, but I can't disclose that sort of information."

- **Forced answer questions** – these are negatively formulated questions. Take your time to think, there is no hurry. Always reformulate the question neutrally, then give an answer: "Is the product unreliable because of A or B?" You: "The reliability of the product. Our product is as reliable as current technology allows. We at Smithson Metals are leaders in research in this field. We have not reached 100% reliability with this product but are definitely working towards that. At the moment we're working on both A and B."

- **Negative comments** – these are not always questions, but can be comments based on previous experience, or stories from elsewhere. Focus on the things you know. You don't need to comment about something you don't know: "... the same product launch in Shanghai was disastrous." You: "Well, I'm afraid I can't comment on that – I'm not involved in it. What I do know is that our product launch is carefully planned for this market here, starting next month."

- **"Obvious" questions** – these are questions about points that you talked about at least ten times in your presentation. Don't say something like "As I said before" or "As I mentioned in the first section" – you don't need to make your audience feel as if its questions were insignificant. For some reason, the questioner didn't get the message. Show any relevant slide again, answer concisely and move on: "The last quarter figures. (*Go back to relevant slide.*) Yes, they're here. The most important point is that they stayed constant when compared to the previous year."

- **Questions already asked** – these are questions that someone has already asked. Don't point out that someone has already asked – never make your audience feel unimportant. For some reason, the questioner didn't hear or understand the answer. Maybe they were thinking about something else. Answer it concisely and move on. If the audience is annoyed with the person asking the same question, then they are annoyed with that person and not with you.

2 Five-step technique

1 Listen
2 Pause
3 Repeat (all or part)
4 Answer
5 Link (if suitable)

You can use this technique for all of the reasons described below.

- Listen – this shouldn't need to be said, but people so often just wait for the others to finish, instead of really listening. Listening can save a lot of time. Don't move around; this could look as if you are not listening. Stand still and just listen.

- Pause – take the time to think of what you want to say (including any important English words). Pausing can feel strange when you first do it, but it creates a very good effect as well as being useful.

- Repeat or reformulate all or part of the question – this is especially useful for two reasons:

  - to be sure you answer the question you were asked,

  - to make sure everyone has a chance to hear the question with larger audiences.

  You may not need this step if you are presenting to a small group when the question is clear and you know that everyone can hear.

- Answer – keep it short and to the point, as soon as you have finished, stop talking. People often continue answering, repeat themselves, go on and on (and on and on) well after the questioner has nodded to indicate that his question has been answered. Keep it short. If people want more details, they will ask. If they want less information, there's not much they can do.

- Link to a main point – if suitable, link to one of your main points to help the audience remember what's important. This is especially useful when answering difficult questions (see previous section for more details). A good technique is to link to what is happening "here and now".

3 Attitude

Your questioner excitedly asks a question, speaking fast and forcefully, almost aggressively. You answer in the same style: fast, forcefully, almost aggressively, too. The next questioner joins in (or the same one again) and is maybe a bit faster, a bit more forceful. Your answer matches the style: a bit faster and a bit more forceful and ... before you know it ... you're in a fight. You may not even be disagreeing with each other!

> If *you* stay calm and answer slowly, clearly but firmly, the situation should not escalate. It's difficult to "fight" with someone who doesn't "fight" back.

Alternatively, your questioner is calm, but asks a question you didn't want. You reply defensively – defending your decision, defending the reasons, instead of stating the reasons calmly and neutrally. The sentence "Yes, it was a difficult decision. At the time, it seemed like the right decision with the information we had then," can sound like a defence or a neutral statement, depending on the speed and tone of voice. If you start to defend yourself, the questioner or other people are more likely to attack, just because you are defending. Again, if *you* stay calm and answer slowly, clearly and strongly, the situation should not escalate. It's not necessary to attack someone who isn't defending himself.

What about really difficult questions, e.g. if the questioner is right with his objection? You should already be prepared for the question, but if not, take the time to think. Is he right? Can you admit it? What *really* happened?

Useful answers

- You're asking why we didn't take option B? Well, back in March, under those conditions, we decided that option A was best. If we had the same situation again today and the same conditions, option A would still look best. Now, with the new information, we know that option B would be better. We couldn't have known that back in March.

- You'd like to know why we didn't do XYZ? Well, I would have liked to have done XYZ, for sure. There are a lot of things I would like to do in addition to what we do now. I think that could be true for a lot of people here today. We didn't do XYZ because we had to make a choice with the resources and manpower we had available, we chose to do RST and UVW.

Watch other people dealing with questions at work, on talk shows or on the news to see how situations can escalate or be de-escalated.

Handling interruptions and disturbances

Helicopters, emergency vehicles, people coming in, phone calls – the list of possible disturbances is endless. One rule is golden: Disturbances take priority. If something else is happening and distracting people's attention, it's difficult for them to listen to you at the same time, and you should act. However, you only need to act when something is really a disturbance – this means it is really distracting people's attention. In some cultures such as those of Italy, Kuwait and Saudi Arabia, it's completely normal to keep cell phones switched on the whole time and often to take the calls in the middle of a presentation, without leaving the room. If they are the customers, then you should accept it. If it's a disturbance to the others, then they will ask the person to leave the room.

As with technical problems (page 55), you should do what you want your audience to do – stay calm and focused. One of the simplest ways of handling interruptions and disturbances is to stop talking and see what happens. As always with your presentation, however, a lot depends on the setting and formality – size of the room, number of people and level of familiarity. A technique that is suitable for one hundred people is not necessarily suitable for a group of three.

If it is not a disturbance to others, then it may be alright to let people talk a bit in the background, now and again. Indeed, it's routine for many polychronic cultures such as the

French, Italians and Spanish. Put yourself in the audience's shoes and ask yourself – do you really need total silence the whole time?

Be very careful not to act like a schoolteacher dealing with children. Remember these are adults, just like you, and sometimes there is a need to talk.

Getting attention at the beginning

Especially with larger groups of people who have come a long way, the amount of noise can be quite intense to start with and may take some time to stop. A louder voice than usual, repetition and stopping mid-sentence is a good way to get attention: "Good morning ladies and gentlemen (*talking*), good morning, (*short pause, talking continues*) ... ladies and gentlemen, good morning ... (*longer pause until talking stops*), ladies and gentlemen. We are here today to ..."

Another option is silence – stand at the front, raise your hands and get eye contact with people who are talking, then bring your hands together and start.

Audience member working on laptop

People working on laptops can create a distraction. If you let them carry on, you might be sending the wrong message to your audience – namely, it's okay to do something else during my presentation. More forcefully expressed – it's alright if you don't listen to me. If it's standard procedure in your company, then a change of policy is necessary before anything can be done. If it's not standard procedure, you can try silence, moving closer to the person(s) or making a comment.

Useful comments

- It would be very helpful if we could all stop what we're doing, then we could get started. Okay ... (*pause*) ... I'd like to get started now (*move closer to the people working on their laptops*).

- I understand that everybody has a lot to do ... (*pause, long if necessary, to get eye contact*), but I'd really like everybody to concentrate fully for the next twenty minutes, because we need to make an important decision based on the information in this presentation.

> Don't use names to highlight the "naughty" people – use "we" or "it".

If none of the above works you could either ignore them or try asking how long they need: "I see some people are busy with e-mails – how long do you need?"

Interruptions from people coming in

People coming in can be distracting in smaller groups (usually less distracting in larger groups). Either just stop talking until the person is settled, or make a short comment. Don't forget, even though you would prefer everyone to be on time, people coming late are absolutely normal. Your best policy is to make them feel welcome – use their name if you know it. If a lot of people come late and it's annoying, then leave it to the audience to make comments.

Useful comments

- (*Person comes in*) "Aha, good morning Giovanni. Come in –
  this seat is free here, or that one over there. You can hang
  your coat up in the corner and there's coffee over on the
  side ... (*wait until most of the noise has subsided*) ... let's
  just wait until you're settled ... okay, where were we?"

- (*Another person comes in*) "Aha, good morning Brian. This
  seat here is still free. You're not the only one who was
  late, don't worry. You can hang your coat up in the corner
  ... (*wait until most of the noise has subsided*) ... okay, let's
  carry on ..."

Audience member on a telephone call

A telephone call can also create a distraction. If necessary,
state clearly but politely how telephone calls should be dealt
with at the beginning of your presentation (see "Introduction
components", page 48). If you are in a small group and
someone takes a call, just stop talking and get eye contact.
The person will either finish the call quickly or leave the
room. If you are in a larger group and someone taking a call
causes a distraction, try the same, or walk up to them in a
friendly manner. If it's someone important, either ignore it
and continue with your presentation or wait for them to
finish.

What to say and what not to say

You probably don't need to say anything, but if so, then be careful not to sound over-polite and sarcastic: "Can you take the call outside please?" is okay, but NOT "Could you possibly take the call outside please?" This can easily sound sarcastic.

Outside noise

Drilling and hammering, window cleaning, whatever – it's not unusual for presentations to take place in meeting rooms in hotels and then be disturbed by background noises or activities. Don't be afraid to take a short break and either telephone reception, go there yourself, or ask someone to go, and ask them to stop or do the work at the end of your meeting or in your lunch break. If the disturbance doesn't cease, name the problem, state the options, decide on the action and carry it out – keep your audience informed at all times.

Useful comments

- Let's wait for that to go by ... good, okay, where was I? Ah yes, one of the key advantages of this system is ...

- We seem to have some workmen nearby. We can either continue or I can go and see if they could stop ... (*Wait for the audience's reaction, if any, then make a clear decision.*) I'd prefer as little background noise as possible, so I'd like to take a two-minute break and just ask reception to tell us what's happening.

Audience members talking

As already mentioned at the beginning of this section, you only need to act if the talking really is a disturbance – so probably not in France, Italy and Spain and to a certain extent, America and Britain. Stopping to talk yourself is usually very effective, moving closer to the people if necessary and finally, if really necessary, making a comment:

Useful comments

- Is there anything unclear at the moment? (*referring to your presentation with your hand*)
- Is there something we should all be talking about?
- I can see you're talking about something important. Is it relevant right now or could you leave it until later please?

> Don't translate from German directly and say "Can we have only one discussion please?" This can sound extremely rude and aggressive.

Unwelcome interruptions from audience

If someone interrupts with something important but not urgent, just remember why you are all there and ask everyone to focus on "here and now". Stay calm and fix when to do the other task(s).

Examples

(*Other:*) We really need to set the time schedule for ABC and DEF. It would be a good time to do it now, because everybody's here.

(*You:*) Yes, everybody is here, but I'd really like to finish this report first. That's the main reason for this meeting today. Can we do it another time?

Or: (You:) Can we do it as soon as I'm finished? Just before lunch? (*Other:*) Well, no, not really, I have another appointment at lunchtime. (*You:*) Okay, how about finishing five minutes before lunch, if that's okay with everyone else?

Heating and light disturbances

If you see your audience yawning, it might not be due to dryness of content but lack of fresh air. Either turn down the heating or open or ask people to open windows as required. Alternatively, you can take a five-minute break and leave doors and windows open.

Useful phrases

- Hmm, I think we need some fresh air. Let's open the windows and take a five-minute break. Can we start again at 11.20 or so? Good.

- Can somebody help me to turn down / off the heating and open the windows?

If someone objects because they like the heat, then just say:

- Okay, then let's leave the heating on and air the room every now and again.

If you have problems with sunlight in people's faces:

- Could somebody lower the blinds a little?
- Could you pull that curtain so that we don't have direct sunlight?"

Ending your presentation

Whether short or long, informal or formal, your presentation needs to have a clear, well-structured and professional ending.

This chapter includes information on:

- making a good finish (page 114),
- ending components (page 116),
- saying goodbye (page 118).

Making a good finish

"Well, okay, that's all we've got time for ... um ... thank you for ...um ... your attention and ... um ... thank you for listening and... um ... thank you and goodbye."

This sort of weak ending is not unusual. Many people put a lot of effort into their presentation introductions and slides, but don't give much thought to the end. You need to end your presentation well because people most often remember your last words best. A good ending is important especially when your audience comes from all over the world and you don't meet very often. You want them to remember you and your message. A good ending is the reverse of a good introduction – opening, objective, overview and organisation – and has four main components as follows:

1 ending signal,
2 review: gives people a last chance to remember any remaining questions,
3 objective achieved: gives value to the time spent,
4 next steps (and closing): provides direction.

You have a lot of different alternatives at the end, depending on if you want (or have time for) any (more) questions. You can have a final discussion and / or question time – depending on the situation, how many questions you already had and the timing. You can review before or after the final discussion, or both, depending on the topic.

International viewpoint

The ending should not be the time to present new information, recommendations or conclusions. It is a very brief review of what has been covered and is likely to be much shorter than a typically correct German ending to a presentation. A typical ending would be only five or six sentences long, lasting a minute or so.

English language points

Be sure to formulate objective(s) achieved carefully – don't directly say "we all now understand." but "I hope we all now have a good idea of what the process will look like". This indirect formulation is more suitable for worldwide use.

Example

 So, we're at the end now. We looked at the new product features, the planned integration and potential challenges – does anybody have any more questions at all? (*People raise hands.*) Yes? ... (*More questions asked.*) Okay, we have time for one more question right now, but you can speak to me in the lunch break or contact me again anytime – so, the last question for now – the gentleman at the back ... Good. Well, I hope that everybody now has a clear picture of the product, how we plan to integrate it and overcome potential challenges. Our next product presentation will be in Madrid in June. If you need more information in the meantime, you know how you can reach me. Thank you very much everybody.

Ending components

Ending signals

- So, that brings us to the end.
- We're at the end now.
- Looking back ...

Review

- We('ve) looked at A, B and C.
- We('ve) covered A, B and C.

Opening discussion

- Does anybody have any (more) questions?
- Are there any (more) questions?

Don't say "Next question, please." This can sound quite military and aggressive.

Closing discussion

No need to say anything like "We're running out of time" or put in the word "only" – "We only have time for one or two more questions" – these are negative comments.

- Okay, we have time for one more question right now ...
- We have time for one or two more questions here today, and then we have to finish ...
- ... but you can speak to me in the lunch break or contact me again anytime.
- ... but I'm here all day and reachable back in the office next week.

Objective(s) achieved

- So, I hope you now know how important A, B and C are for you in your job.
- Well, I hope you now have a clear picture of progress on this project.
- That's all for this morning – I hope you now know how technologically advanced, reliable and flexible our (product) is for you.
- Okay, so, I hope you can now see why we at Alpha Chemicals are the very best partner for you.
- Good. I hope you now have a clear overview of the figures.
- I hope you are now updated on our marketing campaign for the third quarter.

Next steps and contact details

- The next meeting on D, E and F will be organized by our team assistant.
- Our next progress meeting is in ... (*place*) on ... (*date*). The next step is to discuss any further requirements. We look forward to working together with you in the future.
- Our next financial update will be in ... (*place*) on ... (*date*). If you have any points to discuss in the meantime, you know how you can reach me.
- Should you have any further questions in the future, my e-mail is ... (*slide*).
- You can find me in the company staff directory if you ever need any information; just call me or send me a mail.

Thanks and closing

A good strong ending is to say "thank you everybody" and then stop talking. People often say "thank you for ..." then add "your attention and ..." then have to say something else and find it hard to finish because they continue to say "and ...". Thanking people for their attention can sound like school or the army. A good, clear "thank you" is suitable for most occasions.

- Thanks a lot everybody.
- Thank you everybody.
- Ladies and gentlemen, thank you very much.

> Don't say "thank you and goodbye" at the end of a presentation. This is completely wrong and can be misinterpreted as "thank you and go away". "Goodbye" should only be said when people leave each other, physically go away and do not see each other for a time.

Saying goodbye

When you leave someone, at the company reception, the station, the airport, wherever, that is the right time to say goodbye (and not at the end of your presentation). Saying goodbye can often be unnatural and awkward. People want to be polite but don't know exactly what to say. Don't hurry – play "ping pong" by making return comments.

Examples

A: Okay then, thanks a lot for coming and talk to you next week.
B: Yes, we'll be in touch.
A: Alright then, bye for now.
B: Bye for now.

A: I'd like to thank you for coming.
B: It was our pleasure.
A: It was good to meet you.
B: It was good to meet you, too.
A: We'll be in touch about the next steps.
B: Yes, we'll look forward to hearing from you.
A: I hope you have a safe journey.
B: Thank you very much.
A: Oh, please give my regards to Jonathan when you see him.
B: Yes, of course. And thanks once again for inviting us over.
A: It was a pleasure.
B: Okay then, goodbye for now.
A: Goodbye for now.

A: So, here we are.
B: Yes, on behalf of our company, we'd like to thank you very much for the invitation. It was an honour for us to come and visit you and your people and to see the products that you produce. We very much look forward to doing business with you in the future and we hope that we will have a long and fruitful business partnership.
A: Thank you very much. We were delighted that you could come to visit and very much enjoyed getting to know you. We also look forward to a long and fruitful business partnership. We wish you all a safe journey home and we will be in touch in the next few days.
B: Thank you, we look forward to hearing from you.
A: Then, we wish you a safe journey.
B: Thank you, goodbye.
A: Goodbye for now.

B: Goodbye.

International viewpoint

Length and formality of saying goodbye are the two main differing factors worldwide. British and American will tend to be briefer and less formal than Germans. Chinese, Japanese and other Asians will tend to be longer and more formal. It is, however, again very dependent on the situation. Watch people's body language and try to "see" how they feel – if they want to make quick goodbyes or are happy with more formal short speeches.

Useful phrases

- It was nice / good to meet you. *Answer:* It was nice / good to meet you, too.

- It was nice / good meeting you. *Answer:* It was nice / good meeting you, too.

- We'll be in touch. *Answer:* Yes, we'll look forward to hearing from you.

- We'll let you know. *Answer:* Yes, we'll wait to hear from you.)

- Talk to you next Monday, then. *Answer:* Yes, we'll talk next Monday.

- Talk to you soon. *Answer:* Yes, talk to you soon.

Useful examples

Basic outline – non-specific content

Good morning, I'm Benno Donauer from the electronic engineering department at Opus GmbH and it's good to be here in Rio with you on this sunny Friday morning.

By the end of this presentation, you should know why A, B and C are important for you.

First, we'll look at A, then move on to B and finish with C.

We have 40 minutes altogether.

If you have any questions at any time, do please ask.

Okay, let's start.

So, why is A important? This is why ... (*click to next slide*).

What is the current A? Here it is ...

What does the future A look like? This is it ... (*click to next slide*).

Next ... / Now ...

This graph shows A, with A1 on the left and A2 on the right.

This chart shows A3 compared to A4.

So, that brings us to the end.

Are there any (more) questions?

So, we('ve) covered A, B and C.

I hope you now know how important A, B and C are for you.

The next meeting on D, E and F will be organized by our team assistant.

If you have any points to discuss in the meantime, you know how you can reach me.

Thank you everybody.

Product presentation

For those who don't already know me, my name's Bernd Niederhuber, I'm from the Electronics Assembly Systems Department at Drive Technologies AG in Berlin.

So, here we all are in Madrid, today, October 5th, 2008, and by the end of this presentation (*show title slide*) "Serrano Limitido and Drive Technologies", I hope you will know how technologically advanced, reliable and flexible our machines can be for you, and be able to see why we at Drive Technologies are the very best partner for you.

We'll look at the X-machine technical functionality to start, then move on to possible adaptations for you, Serrano Limitido, and finish with our service offer – how we can work well together.

We have around three hours scheduled, with a break halfway through for coffee.

If you have any questions at any time, do please ask me.

Good.

Let's get started.

What are the main technical functions? Here they are ... (*click to next slide*).

Okay, we covered technical functions; now let's look at possible adaptations ...

So, finally, our service offer to you at Serrano Limitido.

We're at the end now. Does anybody have any more questions?

So we looked at the X-machine technical functionality, then possible adaptations for you, Serrano Limitido, and finally our service offer – how we can work well together.

I hope you now know how technologically advanced, reliable and flexible our machines are for you and that you can see why we at Drive Technologies are the very best partner for you.

We look forward to working together with you in the future.

Ladies and gentlemen, thank you very much.

Index

Bibliografische Information der Deutschen Bibliothek
Die Deutsche Bibliothek verzeichnet diese Publikation in der Deutschen Nationalbibliografie; detaillierte bibliografische Daten sind im Internet über http://dnb.ddb.de abrufbar.

ISBN 978-3-448-08734-5
Bestell-Nr. 00972-0001

© 2008, Rudolf Haufe Verlag GmbH & Co. KG, Niederlassung Planegg b. München
Postanschrift: Postfach, 82142 Planegg
Hausanschrift: Fraunhoferstraße 5, 82152 Planegg
Fon (0 89) 8 95 17-0, Fax (0 89) 8 95 17-2 60
E-Mail: online@haufe.de
Internet: www.haufe.de
Redaktion: Jürgen Fischer

Gesamtbetreuung: Sylvia Rein, 81371 München
Lektorat: Nicole Jähnichen, 80333 München; Sylvia Rein, 81371 München
Umschlaggestaltung: Kienle gestaltet, 70178 Stuttgart
Umschlagentwurf: Agentur Buttgereit & Heidenreich, 45721 Haltern am See
Druck: freiburger graphische betriebe, 79108 Freiburg

Die Autorin

Jaquie Mary Thomas

stammt aus Oxford, England. Als Diplom-Sozialpädagogin (FH) ist sie auf Erwachsenenbildung spezialisiert. Sie ist von der University of Cambridge (England) als „Teacher of English as a Foreign Language" zertifiziert und von dem International Cultural Institute (Portland, USA) in „Intercultural Foundations" ausgebildet. Ihr Institut „International Communication Training" berät und schult seit 1992 Mitarbeiter und Führungskräfte von internationalen Konzernen.

Die Schwerpunkte sind Trainings in Business English, TOEFL, GMAT, Präsentation, Moderation, interkultureller Kompetenz, Führung und Teambildung. Darüber hinaus ist sie Mitglied in SIETAR (Berufsverband für interkulturelle Zusammenarbeit und Internationalisierung) und war Lehrbeauftragte an der Ludwigs-Maximilians-Universität München.

Kontakt: www.intcomtra.de

Weitere Literatur

„Englisch für die Personalarbeit" von Thomas Augspurger, Schimon Grossmann u.a., 228 Seiten, mit CD-ROM, € 34,80. ISBN 978-3-448-07489-5, Bestell-Nr. 04267

„Controlling-Fachbegriffe Deutsch-Englisch, Englisch-Deutsch" von Anette Bosewitz, Dr. René Bosewitz, Frank Wörner, 280 Seiten, mit CD-ROM, € 29,80. ISBN 978-3-448-06030-0, Bestell-Nr. 01418

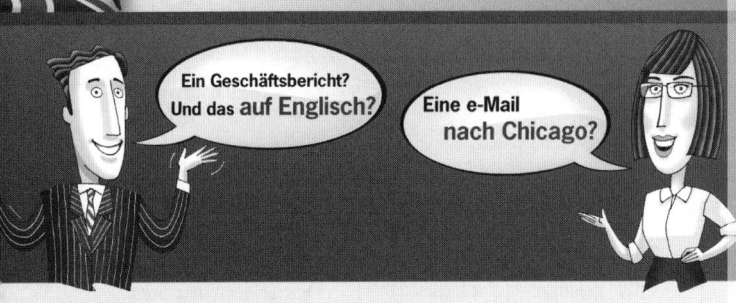

TaschenGuides – Qualität entscheidet

Bereits erschienen:

■ Der Betrieb in Zahlen

- 400 € Mini-Jobs
- Balanced Scorecard
- Betriebswirtschaftliche Formelsammlung
- Bilanzen lesen
- Buchführung
- Businessplan
- BWL Grundwissen
- BWL Kompakt – die 100 wichtigsten Fakten
- Controllinginstrumente
- Deckungsbeitragsrechnung
- Einnahmen-Überschussrechnung
- Finanz- und Liquiditätsplanung
- Die GmbH
- IFRS
- Kaufmännisches Rechnen
- Kennzahlen
- Kleines Lexikon Rechnungswesen
- Kontieren und buchen
- Kostenrechnung
- Kleine mathematische Formelsammlung
- VWL Grundwissen

■ Mitarbeiter führen

- Besprechungen
- Führungstechniken
- Die häufigsten Managementfehler
- Management
- Managementbegriffe
- Mitarbeitergespräche
- Moderation
- Motivation
- Projektmanagement
- Spiele für Workshops und Seminare
- Teams führen

■ Karriere

- Assessment Center
- Existenzgründung
- Ich-AG – mit Gründerzuschuss selbstständig

- Jobsuche und Bewerbung
- Vorstellungsgespräche

■ Geld und Specials

- Die neue Rechtschreibung
- Eher in Rente
- Energieausweis
- IGeL – Medizinische Zusatzleistungen
- Immobilien erwerben
- Immobilienfinanzierung
- Sichere Altersvorsorge
- Geldanlage von A–Z
- Web 2.0
- Zitate für Beruf und Karriere
- Zitate für besondere Anlässe

■ Persönliche Fähigkeiten

- Allgemeinwissen Schnelltest
- Ihre Ausstrahlung
- Business-Knigge – die 100 wichtigsten Benimmregeln
- Mit Druck richtig umgehen
- Emotionale Intelligenz
- Entscheidungen treffen
- Fitness für Beruf und Karriere
- Gedächtnistraining
- Glück!
- IQ-Tests
- Knigge für Beruf und Karriere
- Knigge fürs Ausland
- Kreativitätstechniken
- Manipulationstechniken
- Mind Mapping
- NLP
- Persönliche Situationen meistern
- Schneller lesen
- Selbstmanagement
- Sich durchsetzen
- Soft Skills
- Stress ade
- Verhandeln
- Yoga für Beruf und privat
- Zeitmanagement